COSMOLOGY

SECOND EDITION

Hermann Bondi

With a new Introduction by

Dr. Ian W. Roxburgh

DOVER PUBLICATIONS, INC.
Mineola, New York

Bibliographical Note

This Dover edition, first published in 2010, is an unabridged republication of the 1960 second edition of the work originally published in 1952 by Cambridge University Press, Cambridge. A new introduction has been written for this edition by Ian W. Roxburgh.

Library of Congress Cataloging-in-Publication Data

Bondi, Hermann.
Cosmology / Hermann Bondi.—2nd ed.
p. cm.
"This Dover edition, first published in 2010, is an unabridged republication of the 1960 second edition of the work originally published in 1952 by Cambridge University Press, Cambridge. A new introduction has been written for this edition by Ian W. Roxburgh".
Includes bibliographical references and index.
ISBN-13: 978-0-486-47483-0
ISBN-10: 0-486-47483-6
1. Cosmology. I. Title

QB981.B7 2010
523.1—dc22

2009037039

Manufactured in the United States by Courier Corporation
47483601
www.doverpublications.com

CONTENTS

INTRODUCTION TO THE DOVER EDITION

As a schoolboy there were three books that were primarily responsible for my becoming an astronomer: Eddington's *Nature of the Physical World* (1928), Bondi's Cosmology (1952) and Hoyle's *Frontiers of Astronomy* (1955). The same was true for many of my generation. But of the three it was Bondi's *Cosmology* that I read and re-read as I progressed through University and on to graduate studies.

Bondi was clear about his aims: "to present cosmology as a branch of physics in its own right," and that he did "not regard Cosmology as a minor branch of general relativity or as a branch of philosophy and logic." True to this the book covers basic principles, observations and theoretical models but, fortunately, a fair amount of philosophy permeates the book.

Bondi was best known in Cosmology as one of the creators of the Steady State Theory—but the book is not a polemic on this theory, covering a range of theories that were under debate in the 1950's, the "Big Bang" theory, Eddington's Fundamental Theory, Milne's Kinematic Relativity, McCrea and Milne's Newtonian Cosmology, and the variable constant of Gravity theories of Dirac and Jordan. It is a mine of information for the scholar that wishes to learn of the evolution of ideas on cosmology in the 1930s-60s.

The opening chapters of the book cover the Cosmological Principle, Olbers paradox and Mach's Principle. The Cosmological Principle—at any given time the Universe, on average, appears the same to all observers at rest relative to the mean motion—permeates almost all Cosmology and is discussed in some detail. This was extended to the Perfect Cosmological Principle—the Universe will appear the same at all times—which formed the basis of Bondi and Gold's formulation of the Steady State Theory. Given the observed expansion of the Universe this clearly requires that matter be continuously created so that the mean density is constant in time. Throughout the 1950's and early 60's the Steady State Theory was an equal rival to the Big Bang, but as the data from radio source counts increasingly favoured the Big Bang cosmological picture all but a few dedicated devotees abandoned the Steady State theory. I recall vividly the day when, as a young graduate student, I went to the meeting of the Royal Astronomical Society where the results of the Cambridge radio source counts were to be presented; the London evening newspaper carried the Banner Headline "Bible was right."

Bondi's resolution of Olber's Paradox—that the night sky should be as bright as the surface of a star—was that either the Universe is young or the Universe is expanding. He favoured expansion as this resolved the problem in the Steady State theory but it is now understood that it is the finite lifetime of stars that is the important factor in resolving the paradox. In discussing Mach's Principle—inertial frames and the inertia of bodies are determined by the distribution and motion of the distant astronomical objects—he argues that any theory that assumes an evolving Universe must deal with the possibility that the constant of gravity varies in time, a concept that was subsequently developed by Brans and Dicke into the scalar tensor theory of gravity. Attempts to measure such variation have been unsuccessful with limits on the time scale of variation being longer than the "age of the Universe." However in my view this is a problem worthy of further investigation.

The observational position at the time of writing was considerably different from the situation today—but is of considerable historical interest. The estimated distance to nebulae at that time gave a time scale for the expansion of 1.8 billion years much less than the 4-5 billion year estimates of the age of stars and meteorites. This age problem was much discussed by Bondi, Gold and Hoyle at their regular meetings in Bondi's rooms in Trinity College, and was a major factor in their development of the Steady State Theory. The age problem was resolved by 1958 when Sandage derived a time scale of 13 billion years—close to the present day estimates.

The chapters on Newtonian Cosmology and Kinematic relativity are highlights of the book. In 1934 Milne and McCrea showed that most of the results of Relativistic Cosmology could be reproduced in Newtonian theory, including those with a cosmological constant. This was at a time when the British postal system was at its bcst, Milnc was in Oxford and McCrea in London and they exchanged two letters a day by post—these days you would be lucky to exchange two letters in a week! Kinematic Relativity was developed by Milne in the 1930s and 40s based on a strict application of the cosmological principle, and distance determination by clocks and light signals in special relativity. One of the important contributions was the recognition by Milne that the description of the age and state of motion of the Universe depended on the time scale adopted, and that by a regraduation of clocks to logarithmic time one could describe the uniformly expanding universe as at rest and of infinite age. Milne then went on to try to develop a theory of gravity and dynamics, but although fascinating in showing what can be achieved solely from the application of the cosmological principle, he made little progress in developing physics and his work is now of historical interest.

Bondi's *Cosmology* is a deeply penetrating account of the state of cosmology in the 1950s and was a major inspiration to a generation of astronomers. Some of the ideas discussed remain valid today, others have been discarded or disproved with the passage of time, but the deep insights and foundational issues considered make it a classic and a valuable source of historical information and of ideas.

Ian W. Roxburgh
Research Professor in Astronomy
Queen Mary, University of London

PREFACE TO THE FIRST EDITION

The aim of this book is to present cosmology as a branch of physics in its own right. For this purpose a large part of the book has been devoted to the observational aspect. A coherent representation has been given of all the observational evidence likely to be of cosmological interest.

The part of the book dealing with cosmological theories is necessarily lengthy. It seemed wrong to me to omit any of the major theories, although in my opinion the steady-state theory agrees best with observation and has the simplest and most logical basis. Newtonian cosmology, relativistic cosmology and kinematic relativity have all made such great contributions towards the development of cosmological thought that each of them had to be discussed in some detail. Limitations of space made it necessary to give only brief descriptions of these theories, but the existence of a number of monographs makes further study comparatively easy.

My view on the steady-state theory has already been mentioned. It may also be in place to state here that I do not regard cosmology as a minor branch of general relativity or as a branch of philosophy and logic. If this book can make the reasons for my attitude clear it will have served a useful purpose.

My thanks are due to Mr A. R. Curtis, Mr T. Gold and Dr R. A. Lyttleton for very helpful criticism.

I also wish to acknowledge gratefully the work of the Cambridge University Press.

H. B.

CAMBRIDGE
27 October 1950

PREFACE TO THE SECOND EDITION

Since the first edition of this book went to press in 1951 considerable changes have occurred in cosmology. They are in the main due to the drastic revision of the time scale by Baade and Sandage, which has largely resolved the time-scale problem so prominent eight years ago. No major independent development has occurred in theoretical cosmology, but readjustment of the various theories to this new situation is taking place. The work is substantially complete for Lemaître's model and for the steady-state theory, but a great deal remains to be done for the rest of relativistic cosmology and the other theories.

Another development of great interest in cosmology is the solution of the problem of nucleo-genesis which has changed attitudes concerning point-source models. Radio astronomy is likely to play an important part in the future in spite of the mistaken nature of early claims.

The new edition has attempted to take full account of all these developments. It may be regarded as complete in this respect but, as for technical reasons wholesale deletion has been impossible, certain sections, especially those on number counts, now of primarily historical interest, are still included. Nevertheless the book aims to give as up-to-date an account of cosmology as the first edition did eight years ago.

The author is deeply indebted to Dr R. A. Lyttleton for seeing the book through the press.

H. B.

LONDON
March 1959

PART 1

PRINCIPLES OF COSMOLOGY

CHAPTER I

PHYSICS AND COSMOLOGY

1.1. It is customary to devote the first few sentences of a scientific text-book to the definition of the range of the subject and a brief discussion of its limits. In cosmology such a definition deserves more than brief attention, since many of the differences of opinion so characteristic of the subject may be traced back to differing conceptions of its boundaries and its place in physics and astronomy. In particular, there are two important approaches to the subject so different from each other that it is hardly surprising that they lead to different answers.

The development of dynamics (and, later, of other branches of laboratory physics) is a story of success following success, and it is not surprising that each new success led to new extrapolations. The 'falling-apple' law of gravitation was readily applied to the moon, the identification of spectral lines in the laboratory was immediately extended to stellar spectra. It is natural that after many such extrapolations had been successfully applied to observations of more and more distant objects two questions should have been asked: 'What is the largest set of objects to which our physical laws can be applied consistently and successfully?' and 'What is the largest set of all physically significant objects?' Either of these sets was called the universe without, at first, any attempt being made to distinguish between the concepts. The distinction, which becomes important only at a later stage, will be discussed then.

One possible mode of progress in answering these questions was to construct mentally a number of different models of such a universe and to investigate whether the physical laws could be applied to them consistently. The first such investigation seems to have been due to Olbers (1826) and will be discussed in Chapter III. Many other such investigations have been carried out since. It is not surprising that such a mode of attack led to a considerable number of model universes, each of them interesting and remarkable in its own right, and the question which of them was the 'actual' universe became of lesser interest. According to this point of view cosmology is one of the many testing grounds of physics where the laws of nature, as known from terrestrial experiments, are applied to the construction of a range of possible 'universes'. Which one of these possible 'universes' is the one actually realized then becomes a secondary question, the answer to which must at present be found observationally, since there are no theoretical grounds for preferring any one of the possible models to any of the others.*

According to this school of thought, then, cosmology is the largest workshop in which we may assemble equipment, the elements of which are entirely composed of terrestrially verified laws of physics. All the models constructed are of interest, though one may be particularly exciting in being a picture of the actual universe in which we live. Should none of the simpler models constructed correspond to the observations made of the actual universe, then this would be mildly perturbing in forcing us to construct more complicated ones, but would not invalidate the assumptions of the theory.

The opposite point of view is reached from investigations in the borderland between physics and philosophy. The dislike of many nineteenth-century physicists for philosophy is presumably the reason why this line of approach was largely neglected until recently.

There are two such lines of argument leading to similar conclusions. When the fundamental assumptions of physical science are investigated it is found that they are illogical and untenable unless certain properties are assigned to space and time, and these

* Three quotations may be given characterizing this point of view:

'For if, as has been held throughout our previous discussions, the whole dynamical world...can be considered without regard to existence...can be applied to universes which do not exist....This being so the universe is given, as an entity, not only twice, but as many times as there are possible distributions of matter....' (B. Russell, *Principles of Mathematics*, 2nd ed., p. 493.)

'General relativity, on the other hand, investigates as many as possible of the patterns to which Nature might conform and leaves to observation the determination of the pattern actually realized.' (G. C. McVittie, *Observatory*, **63**, 281, 1940.)

'It is no part of the relativistic theory of cosmogony to assume a uniform large-scale distribution of matter throughout the universe. As to that we are content to accept whatever data astronomical observations may provide.' (A. S. Eddington, *Sci. Progr.* **34**, 227, 1939.)

in turn are equivalent to certain suppositions concerning the world at large, the universe. In this way the fundamental laws of cosmology are immediate consequences of the *a priori* assumptions involved in any physical science. These fundamental laws play the part of axioms, and the main body of cosmology may be deduced from them in the manner of a discipline of pure mathematics. It may be found necessary, in one or two places, to supplement these basic axioms by other, less fundamental ones, in order to make progress, and in the choice of these supplementary axioms terrestrial physics will be a guide. But it should be possible to reach, in a deductive manner, conclusions of importance not only to cosmology but to all physics. According to this point of view, then, cosmology is the most fundamental of the physical sciences, the proper starting-point of all scientific considerations. It may be true that we still have to rely in one or two places on terrestrially obtained knowledge, but these gaps in the deductive chain may (it is hoped) be filled in soon, making cosmology the truly all-embracing science, the long-sought link between philosophy and the physical world.

The other philosophical argument, while rather more vague, leads to the same result by suggesting that the world should be simple when viewed on a sufficiently large scale. This is the natural complement to the underlying hope of all atomic research, that matter should be simple when viewed on a sufficiently small scale. This deep feeling that complexity is a feature of phenomena on our own scale only is one of the most powerful driving forces of science, which always expects to find the simple at the root of the complicated.

The contrast between the 'extrapolating' and the 'deductive' attitudes to cosmology is very great indeed, and most of the workers in the field adopt an intermediate position combining parts of both. However, it is clear that there is room for wide divergences of opinion. This in itself is no disadvantage, since serious progress is impossible in a state of self-satisfied unanimity. There is, however, a widespread lack of understanding and of appreciation of the differing points of view that makes the disagreements less fruitful for advancing the subject than they would otherwise be. In particular, the tendency on the part of the empirical school to regard any work of the deductive type as being 'too speculative' seems to be based on a complete misunderstanding of the scientific scale of values. The overriding principle must be that of the economy of hypotheses, but in comparing different theories according to this principle, one must take account of *all hypotheses* involved in them whether originally tacitly assumed or not. Replacing an old assumption with which many are acquainted by a new one (even if strikingly novel) does *not increase* the number of hypotheses required, and it is quite wrong to consider such a change a disadvantage. The value of a hypothesis depends primarily on its fruitfulness, i.e. on the number and the significance of the deductions that can be made from it, and not on whether it requires a change in outlook and is considered 'upsetting'.

There seems to be a widespread tendency to consider any extrapolation from observational data, however great, to be 'self-evident' and therefore not as a special hypothesis to be counted in computing the number of hypotheses required by the theory. This attitude, understandable though it is, is clearly utterly mistaken. When, for example, Milne (1935) suggested that the ratio of the speeds of dynamical and atomic phenomena varied with time in a certain slow but significant manner, he was accused of having introduced an additional hypothesis. This is plainly not so: he merely *replaced* the customary hypothesis that the ratio of the speeds was the same at all times by another hypothesis, viz. that the ratio varied in a certain manner. His assumption, though new, was no more speculative than the old one. Therefore the old and the new have to be treated on the same footing. Hence both of them have to undergo equally the tests as to whether they are fruitful, i.e. whether conclusions can be drawn without additional assumptions, whether these conclusions form a self-consistent scheme, and whether this scheme agrees with observation.

It is a dangerous habit of the human mind to generalize and to extrapolate without noticing that it is doing so. The physicist should therefore attempt to counter this habit by unceasing vigilance in order to detect any such extrapolation. Most of the great advances in physics have been concerned with showing up the fallacy of such extrapolations, which were supposed to be so self-evident that they were not considered hypotheses. These extrapolations constitute a far greater danger to the progress of physics than so-called speculation.

On the other hand, there certainly exists a type of speculative theory which is useless and is rightly widely condemned. In it *ad hoc* assumptions follow each other in rapid succession, each trying to cover a separate piece of ground. This situation is apt to arise if the initial assumptions of the theory were not sufficiently powerful to allow any far-reaching deductions to be made without numerous further hypotheses. The mere number of special assumptions made clearly rules out serious consideration of such a theory.

This condemnation is, however, frequently extended quite incorrectly to cover a very different case. A theory may be based on a powerful hypothesis which for various reasons may be widely regarded as implausible. The theory may then very properly attempt to explore *all* the consequences of this powerful hypothesis however remote from its starting-point. This is not only the correct thing to do but is most desirable. The initial emotional judgement of the likelihood of the truth of the basic hypothesis is completely irrelevant; what is important is whether its consequences form a self-consistent scheme agreeing with the actual observations, though not necessarily with commonly accepted extrapolations from the observations. Nevertheless, this important exploration of the remote consequences of such an assumption is frequently thoughtlessly condemned as 'heaping speculation upon speculation', surely a most inappropriate term. The logical examination of *all* the results of a hypothesis is not only legitimate, but is the very essence of science.

Just as some adherents of the 'empirical' school tend to regard cosmology as a testing ground for their extrapolations and as a legitimate playground for the geometers, so some adherents of the deductive approach appear to regard cosmology as a purely logical subject. To them all that is of interest in a theory is its logical character, not its relevance to the interpretation of observational data. They are willing to discard a theory only if a logical internal inconsistency is found, and they consider any disagreement with observation as of minor importance only, relating merely to the applicability of the theory.

Such an approach (exemplified by the cosmology of Parmenides, who considered the universe to be a perfect sphere always similar to itself) seems to treat cosmology very differently from the rest of physics and does not treat the observational material as of great importance. This attitude runs counter to the whole spirit of modern science.

A further danger in this outlook is that however much one may try to formulate a purely logical theory, one's knowledge of physics always intervenes, directing, formulating, or at least guiding, the logical development. The extreme difficulty of disentangling this implicit occurrence of physics from logic is alone a good reason for preferring to make progress along explicitly physical lines.

1.2. It is well known that the special theory of relativity owes its existence to the fact that there is an upper limit to velocities of motion and signalling in nature, viz. the velocity of light. This fact has also very important consequences for the subject of cosmology, since it makes it impossible to investigate vast spaces without also contemplating enormous lengths of time. In particular, any knowledge that we may possess about very distant objects must be due to processes (signals) originating on those objects a very long time ago. Although the very definitions of distances and times become ambiguous in this context it is clear that the great intervals occurring raise new questions. If (as is widely held to be true) the universe is in a state of evolution, then light from distant parts of space must also convey information about the early history of the universe. If, on the other hand, there is no such evolution, then the light must contain information about its absence. Therefore the entire question of the origin and the evolution of the universe becomes part of the general cosmological problem. In particular, the nature of the very beginning (if there was one) or, alternatively, the origin of all matter and radiation has formed a specially controversial subject generally termed the problem of 'creation'.

In this context, too, we encounter the habit of the human mind to imagine that complexity is only ephemeral, that the further we dip into the past the simpler the state we shall find. How far this process can be continued is a question to which no agreed answer is in sight. What type of 'last answer' is found most satisfactory depends on individual taste. Hence here less sure guidance is given than elsewhere in science by how satisfying or otherwise a theory is. Broadly speaking three types of answer as to the nature of the 'beginning' have been given, and opinions differ widely as to the relative merits of these:

(i) The 'beginning' is a singular point on the border of the realm of physical science. Any question which refers to antecedents of the beginning or its nature can no longer be answered by physics, and is not a proper question for it.

(ii) The 'beginning' was a particularly simple state, the simplest, most harmonious and most permanent we can imagine. It contained within itself, though, the seeds of growth and evolution which at some indefinite moment started off a chain of complicated processes which by now have changed this to our present universe.

(iii) There was no 'beginning'. The universe on the large scale is either unchanging or possibly going through cyclical changes, but is of infinite age.

It will be seen later how these different answers arise and what their natures are. For the moment it will be sufficient to point out that a theory must at least lead up to this, the problem of 'creation', and that opinions differ as to what constitutes a satisfactory answer.

1.3. A difficulty peculiar to cosmology is the uniqueness of the object of its study, the universe. In physics we are accustomed to distinguish between the accidental and the essential aspects of a phenomenon by comparing it with other similar phenomena. The laws of free fall were found as the common part of numerous experiments on falling bodies, the laws of planetary motion were inferred from a knowledge of the orbits of many planets. This method of abstraction is so customary in physics that it has become usual to formulate physical laws as differential equations (representing the common elements of many phenomena), the solutions of which are only specified by the initial conditions (giving latitude to insert the individual characteristics of each phenomenon). The uniqueness of the actual universe makes it impossible to distinguish, on purely observational grounds, between its general and its peculiar features even if such a distinction were logically tenable. It is this impossibility of direct abstraction from the observations that rules out the usual inductive approach. It is owing to this fact that we are forced to choose between the axiomatic deductive line of thought and the large-scale extrapolation of terrestrial physics as outlined above. In either case we select the important (as opposed to the 'accidental') features of the actual universe by their relation to the theory chosen rather than by any independent criterion.

1.4. The last point that must be discussed in this introduction is what is meant by the 'universe'. Accepting the postulate that the velocity of light is the maximum velocity at which influences can be propagated, then it is evident that all events on or within our past light cone may affect us, and they certainly must be included within our definition of the universe. If we extend this to include all events which may affect us at some time in the future and all events which have been or will be affected by us, then we arrive at the largest set of events that can be considered to be physically linked to us. It is this set that is usually considered to constitute the universe. Some authors, however, consider a different set, viz. the largest set to which our physical laws (extrapolated in some manner or other) can be applied. This set may include points that have never been and will never be physically linked to us and may exclude points which are so linked. The physical significance of this 'universe' is therefore not very clear, but it should be remembered that even with the definition adopted above the actually possible observations concern only a very limited part of the universe.

CHAPTER II

THE COSMOLOGICAL PRINCIPLE

2.1. In spite of the differences in outlook of the various theories of cosmology they all agree in postulating the validity of the so-called 'cosmological principle' which, broadly speaking, states that the universe presents the same aspect from every point except for local irregularities. Although there are wide divergencies of view as to the significance, the necessity, and the logical position of this postulate, the agreement as to its validity is very remarkable, and its utility is beyond doubt.

How does this principle attain its pre-eminent position in cosmology? A great variety of arguments for its validity have been given, and from these we shall select three to represent the most widely different points of view.

The first argument is based on an examination of the fundamental assumption of all physical science—the indefinite repeatability of experiments. According to this assumption, which has grown so familiar that we hardly realize its significance, a repetition of the conditions of an experiment (supposing this to be attainable) will always lead to a repetition of the result. This may, of course, be a statistical result. In a laboratory experiment all the conditions of the experiment are usually under general control except for two—the time and the place of performing it. Repetition implies that the second experiment takes place later than the first, and the position of the laboratory will be altered by the Earth's motion through space, quite apart from the fact that we assume as a matter of course that the experiment is capable of being repeated in another laboratory situated at a different part of the Earth's surface. It follows that, unless we postulate that position in space and time is irrelevant, no conclusions whatever can be drawn from the repeatability principle. For otherwise the repetition of an experiment becomes quite impossible since the condition of space and time cannot be repeated. We see therefore that in all our physics we have presupposed a certain uniformity of space and time; we have assumed that we live in a world that is homogeneous at least as far as the laws of nature are concerned. Hence the underlying axiom of our physics makes certain demands on the structure of the universe; it requires a cosmological uniformity.

In this form the argument is widely admitted to be valid but is still rather vague, and its direct cosmological demands seem at first sight to be rather modest. If physics depends on space and time this dependence must not show itself in periods of time of a few thousand years nor in distances such as are traversed by the Earth in such periods. However, as soon as we leave the strictly laboratory type of physics and attempt to apply our scientific knowledge to distant astronomical objects and to the ancient geological history of the Earth and solar system, the argument becomes at the same time more significant and more questionable. *If* we are still able to apply our terrestrially gained knowledge in such far-flung fields, then the uniformity of the cosmos must be very great. But it is not universally agreed that our knowledge can be directly used in such fields; it is, in fact, maintained by some schools of cosmology that in interpreting observations of far-distant or very ancient objects we must take account of the non-uniformity of the background in some way, to be decided by other considerations. Another view, however, rejects any such attempt to discuss the consequences of a real non-uniformity of the universe.

For in any theory which contemplates a changing universe, explicit and implicit assumptions must be made about the interactions between distant matter and local physical laws. These assumptions are necessarily of a highly arbitrary nature, and progress on such a basis can only be indefinite and uncertain. It may, however, be questioned whether such speculation is required. If the uniformity of the universe is sufficiently great none of these difficulties arises. The assumption that this is so is known as the perfect cosmological principle. It was introduced by Bondi and Gold (1948) in the form of the statement, that, apart from local irregularities the universe presents the same aspect from any place at any time. It was shown by Bondi and Gold that this single principle forms a sufficient basis for developing without ambiguity a cosmological theory capable of making definite and far-reaching physical statements agreeing with observation. This theory is the steady-state theory which will be described in Chapter XII.

The question of the interaction between distant matter and local physical laws is clearly relevant to this argument, and the evidence relating to this problem will be presented in Part 2 of this book.

The chain of reasoning leading to the perfect cosmological principle is emphasized by the steady-state theory but is not taken into account in others. They are based on other arguments which lead to the cosmological principle in a narrower sense, in which all positions in space are regarded as equivalent, but in which variation in time is allowed. First amongst these arguments is what may be called the Copernican principle—that the Earth is not in a central, specially favoured position. This principle has become accepted by all men of science, and it is only a small step from this principle to the statement that the Earth is in a *typical position*, and a wide interpretation of the word 'typical' renders this statement equivalent to the 'narrow' cosmological principle. Alternatively, we may derive the cosmological principle without extrapolation from the Copernican principle if we utilize the observational evidence that the universe around us seems to exhibit spherical symmetry. It may easily be seen intuitively (and has been proved rigorously by Walker (1944)) that since a non-uniform universe can present a spherically symmetrical aspect only to an observer in a special position, homogeneity follows by Copernicus' postulate from local isotropy.

As a final argument we may consider the simplicity postulate as a basis for the cosmological principle. This is the point of view adopted by the proponents of general relativity and is effectively that in our present ignorance of the universe progress may most easily be made by assuming, purely as a working hypothesis, that the large-scale structure of the universe is as simple as possible and that hence it is uniform.

Other arguments also exist (such as the expectation mentioned in Chapter I that objects on a large scale should be simple), and we can conclude that it is now almost universally believed that the universe, on a large scale, is homogeneous, but the strength with which this belief is held varies between the opinion which considers it as fundamental to all cosmological thought down to the opinion that it should be believed until we find out more.

Only the steady-state theory considers the corresponding time-like condition, that the universe should be stationary,* as also being true; all other theories believe that there is a universal 'cosmic' time of physical significance marking off stages in the evolution of the universe and possibly also influencing local physical development.

A related question briefly mentioned above is that of the spatial isotropy of the universe, i.e. of the equivalence of different spatial directions. Homogeneity by no means implies isotropy, although universal isotropy implies homogeneity. For this reason some workers consider isotropy to be a good mathematical starting-point for cosmology, but few seem to consider it directly deducible from physics. However, observational evidence and the simplicity postulate, as well as some other arguments (cf. Chapter IV), combine to make spatial isotropy a widely accepted hypothesis.

It is a curious fact that, in spite of general agreement on the validity of a cosmological principle, its precise formulation is a matter of controversy. The extremely marked local inhomogeneities make it clear that the application of the principle must be confined to large-scale phenomena. But whether it postulates an abstract homogeneous background against which all local phenomena must be measured, whether it is merely an approximation or whether it implies some statements about limits taken over arbitrarily large regions, is still a matter of opinion.

It should be mentioned here that there is a possibility of satisfying the cosmological principle in the following way: It can be imagined that, just as stars are grouped into nebulae, and nebulae into clusters, so clusters are grouped into super-clusters of the first order, super-clusters of the first order into super-clusters of the second order, and so on without limit. In this 'hierarchical' world order it is assumed that the volume per cluster increases more quickly with order than the mass per cluster. Accordingly the mean densities of matter and of radiation

* It should be clearly understood that 'stationary' in this connexion only implies that the general aspect of the universe is unchanging and does not signify an absence of large-scale motions.

vanish in such a model, although the model is homogeneous in a certain sense.

This type of world model was favoured at one time, but at present the general preference is for the simpler type of homogeneous model with finite density, in which all regions are similar to that actually surveyed. The observations appear to support this type of homogeneity (pp. 40, 41).

PART 2

OBSERVATIONAL EVIDENCE

This part forms an attempt to present all the information that may possibly be of significance for the formulation of cosmological theories. Some schools of thought reject a large fraction of the material here presented as of no importance for cosmology, at least at its present stage, but here the question of relevance is left to the judgement of the reader.

In the discussion of the observational evidence the cosmological principle is used. This procedure seems to be justified owing to the wide agreement on the validity of the cosmological principle and also because its absence would make any interpretation almost impossible.

CHAPTER III

THE BACKGROUND LIGHT OF THE SKY

3.1. One of the most marked features of our physical surroundings is the extraordinary paucity of radiation. We have become so accustomed to this fact that we are likely to overlook its significance. We are quite used to the fact that the sky is very dark except where a near star sheds its light. Even in the most speculative theories we hardly ever consider temperatures of 10^{14} degrees at which the energy of the random motion of matter equals its rest energy. We know that the average energy density of radiation in the observed parts of space is a minute fraction of the energy density of matter. We consider the ground state of an atom to be its 'natural' state to which it will return after every excitation by getting rid of its surplus energy in the form of radiation, which space will swallow up gratefully and apparently without question of return. In many ways we consider radiative energy to be 'lost' energy, picturing somehow space as an infinite receptacle, an almost perfect sink. The heating of a star by random radiation travelling around space is so utterly negligible as never to receive mention. All these facts are 'obvious' and are well known as symptoms of the extreme thermodynamic disequilibrium of the observed parts of the universe. The deep significance of this state of affairs for cosmology was pointed out as early as 1826 in a remarkable paper by Olbers.

3.2. Olbers assumed without question that space was Euclidean and that the average number of stars per unit volume and the average luminosity of each star were constant throughout space and time, provided these averages were taken over sufficiently large regions.*

* He assumed that these averages did not vanish, corresponding to the usual formulation of the cosmological principle. A hierarchical world model (p. 4) may lead to zero averages, and then the rest of the argument of this chapter becomes invalid. However, in common with most authors, we shall disregard the possibility of this rather complex model. Another possibility that will be neglected is that the apparent darkness of the night sky is not a permanent feature but confined to our time. Whether in that case the arguments of this chapter become invalid or not does not appear to have been investigated.

These assumptions (the classical equivalent of the perfect cosmological principle) were so natural to him that he did not even discuss them. He similarly assumed that there were no systematic large-scale motions, so that on the average the relative velocity of any two stars vanished. He then showed by a simple argument that these extremely plausible assumptions led to a contradiction, now to be explained.

For consider a large spherical shell of arbitrary centre and inner radius r and thickness dr, where r is much greater than dr. The volume $4\pi r^2 dr$ of the shell should be sufficiently great for the light emitted by stars in it to be approximately $4\pi r^2 dr U$, where U is the product of the average number of stars per unit volume and the average luminosity per star. The intensity of light at the centre of the shell due to stars in the shell is therefore $U dr$, and hence is independent of the radius of the shell. If we surround our shell by a succession of other shells of equal thickness concentric with the first, the outer boundary of each shell being the inner boundary of the next, then each shell will make the same contribution to the radiation density at the centre. Since we can add shells without limit, it would seem to follow that the radiation density at the centre is infinite. We have, however, omitted to account for the possibility that light from a star has been intercepted on its way by another star, and when this factor is included it can be shown that the radiation density at the centre must equal the average radiation density at the surface of a star. Since the centre was an arbitrarily selected point the result must hold everywhere.

The same conclusion may be reached by considering the fact that a static system of infinite age must have reached thermodynamic equilibrium, and hence each star must be absorbing as much radiation as it emits. The same result follows, although the nature of the assumptions is not made quite as clear as in Olbers' proof.

It is important to realize fully just how the cosmological principle is used. We require only that the *average* production of light per unit volume should be constant. It is immaterial to how large a region we must go before we can take averages; it is only necessary that the average should tend to a uniform limit as the volume tends to infinity.

A second interesting point is that the bulk of this enormous amount of radiation arrives from very distant parts, in fact, half from regions so distant that the light has only a 50 % chance of arriving without having been absorbed by other stars on the journey. Using a value for the likelihood of such absorption obtained from observations of our neighbourhood, it turns out that half of this radiation should be due to stars more than 10^{20} light-years distant.

Olbers was greatly puzzled by the absurd result that followed with such strict logic from his simple and natural assumption. He tried to explain away his result by postulating the existence of a tenuous gas absorbing the radiation in transit over very long distances, but far too thin to be detected by the astronomical observations of his day. This 'explanation' will not stand serious investigation. What happens to the energy absorbed by the gas? It clearly must heat the gas until it reaches such a temperature that it radiates as much as it receives, and hence it will not reduce the average density of radiation.

3.3. Since no satisfactory explanation of this type has been discovered and the rigour of the deductive argument employed appears to be unimpeachable, we must conclude that some of Olbers' assumptions are wrong. These assumptions may be restated here as:

(i) The average density of stars and their average luminosity do not vary throughout space.

(ii) The same quantities do not vary with time.

(iii) There are no large systematic movements of the stars.

(iv) Space is Euclidean.

(v) The known laws of physics apply.

We first note that (i) and (ii) are not required in their full form for the deduction of Olbers' paradox. It would be sufficient if the quantities mentioned were constant throughout only those parts of space-time from which light reaches us now. In other words, it would be sufficient if the average luminosity per unit volume that existed one million years ago in regions one million light-years away equalled the average luminosity per unit volume two million years ago in regions two million light-years away, and so on. Any such arrangement that does not, at the same time, satisfy (i) and

(ii) would, however, ascribe an exceptional position to the Earth, and may hence be ruled out by the Copernican principle that the Earth is not the centre of the universe. Hence on assuming Copernicus' principle, (i) and (ii) are necessary conditions for Olbers' paradox.

On the other hand, it can easily be shown that if (i), (ii), (iii) and (v) are assumed the paradox can be derived without using (iv) provided only we assume space to be homogeneous, which seems to be almost a corollary of (i). For then the non-Euclidean nature of space must still be uniform and can only express itself in the form that the surface of a sphere of radius r drawn round any point is always some function of r, say $f(r)$, rather than $4\pi r^2$. The total luminosity of the stars in a shell will then be $Uf(r)\,dr$ instead of $U4\pi r^2\,dr$, but similarly the light arriving from them at the centre will be reduced by a factor $f(r)$ rather than $4\pi r^2$ and will still be $U\,dr$, just as before. The paradox follows, even if we cannot take arbitrarily large r owing to space 'closing up on itself'. For then we would be receiving light from the previous periods of the universe, rather as great circles on a sphere can be continued arbitrarily far, although the area of the sphere is limited.

Our result is therefore that postulates (i), (ii), (iii) and (v) cannot be reconciled with observation and cannot all be valid. Which of them is to be dropped? All theories of cosmology seem to consider (i) to be valid as a straightforward consequence of the ordinary cosmological principle. On the other hand, hardly any theory that attempts to represent our actual universe considers (iii) to be valid, since observations of distant astronomical objects seem to indicate that there are large-scale motions often referred to as the expansion of the universe (cf. Chapter VIII). Many theories do not consider (ii) to be valid, or, indeed, compatible with observations, but the steady-state theory (Chapter XII) firmly bases itself on the strict validity of (i) and (ii). Here we shall discuss only the question of how abandoning either (ii) or (iii) can resolve Olbers' paradox.

If (ii) is not considered valid, then U, the average luminosity per unit volume, may be a function of the time. (It will be shown in Chapter VIII that if the cosmological principle holds in the narrow sense, i.e. for spatial variations, but (ii) is invalid, then there must be a well-defined universal time.) Now as we are looking into space we are looking into the past, and if $U(t)$ was sufficiently small in the distant past the distant regions would not contribute materially to the radiation density. It is easily seen that the mean density of radiation in a static universe (condition (iii) valid) with $U(t)$ variable would be equal to the time integral of $U(t)$ over all past time up to the moment of observation. We conclude, therefore, that Olbers' paradox does not arise in a static universe in which, roughly speaking, the stars did not start to radiate until some moment which can be determined (by observation of the present value of U and the present density of radiation) to have been between 10^8 and 10^{12} years ago.

We next proceed to consider how Olbers' paradox can be resolved when (ii) is retained but (iii) is dropped. In this case we are considering a universe which presents an unchanging aspect (is stationary) but not static. Such a universe may be likened to a river that is in a steady state and hence always looks the same although the particles making up the river are moving. How such a state can exist and what its properties are will be discussed in Chapter XII. The point which we want to clarify at this stage is merely how, in such a universe, the radiation density can be low, although U, the luminosity per unit volume, is constant. This clearly implies that the contribution to the radiation density of light from distant stars is reduced far below Olbers' estimate. In the type of universe we are now examining this can be due only to one cause known to physics: the Doppler shift of light. If distant stars are receding rapidly the light emitted by them will appear reddened on reception and hence will have lost part of its energy. If the recessional velocity of distant stars is great enough the loss of energy may be sufficient to reduce the radiation density to the observed level. The exact mechanism of this process and the requisite motions (often called 'the expansion of the universe') will be discussed later, but it will be seen that such a process offers at least some hope of avoiding Olbers' paradox.

We may then summarize our findings by stating that the observed low radiation density together with the assumption of the narrow cosmological principle shows that, assuming (v) to hold, at least one of the following two statements is true: (*a*) *The universe is young* or (*b*) *The universe is expanding*.

Of course all these conclusions become invalid once condition (v) is dropped. However, the number of possibilities which arise then is so very large and so little guidance is given in choosing any of them that this hardly seems a very scientific approach. A purely *ad hoc* assumption of this type is, however, avoided in the so-called τ-scale formulation of kinematic relativity (cf. Chapter XI). There it appears simply that the frequency of photons diminishes with time in a sort of Doppler shift proportional not to velocity but to lapse of time. This has precisely the same effect on the thermodynamic considerations as a Doppler shift, and is in any case merely a different representation of such a shift.

3.4. The really remarkable thing about the result stated above is that so interesting and significant a conclusion can be reached from such simple observations as the darkness of the night sky when compared to the surface brightness of a star, together with the cosmological principle. It demonstrates how powerful this principle is and shows at the same time how much more powerful still the perfect cosmological principle is, which leads necessarily to statement (*b*) above, since it denies the possibility of (*a*).

The reason for the cosmological significance of such a simple fact as the darkness of the night sky is that this is one of the phenomena that depend critically on circumstances very far away. In laboratory experiments we always deal with quantities that we can influence, so as to examine the dependence of one measurement on another. Thereby we automatically exclude phenomena that depend only, or at least largely, on very distant matter (which is out of our control). In cosmology these are just the observations from which we can derive most information. Difficulty arises only because owing to their constancy and uniformity these effects of long-range influences do not attract our attention very much. We tend to ascribe an obvious and 'absolute' character to such effects (when we notice them at all). The consistent interpretation of such effects forms one of the most important and significant parts of cosmology. Unfortunately, this question of interpretation is also very difficult since we are dealing with uncontrollable and usually constant quantities. Hence the checking of a prediction, which usually forms such a vital link in the formulation of physical theories, does not occur in this field, and we have to rely on less objective and certain criteria, such as how satisfying and how simple a theory is.

3·5. We may at this point consider the question of the thermodynamic state of the universe already briefly referred to above. The present state of the universe as observed is evidently one of extreme thermodynamic disequilibrium. Material is on the whole emitting very much more radiation than it is absorbing, the ratio of the maximum to the minimum temperature of matter is very great and the effective temperature of empty space is near the minimum temperature of matter. In what type of universe can such a state of affairs exist even if only temporarily?

The discussion of this question is rather difficult owing to a peculiar feature of an 'expanding' universe. Since we can be reasonably certain from various observations (cf. early parts of this section, also Chapter v) that our universe is expanding, we must devote some time to this consideration.

We know from our local thermodynamics that there are forms of energy in which it is useless and incapable of further transformation although demonstrably in existence. Energy turned into such a state is therefore effectively lost. Now there is such a form of energy in an expanding universe which is not normally considered as 'lost', but its effective 'loss' is of great cosmological importance.

In an expanding universe light emitted from a source is usually received by matter moving away from the source. The average relative velocity of recession depends on the rate of expansion and on the opacity of the universe but is known observationally to be a substantial part of the velocity of light. Accordingly, the frequency (and hence the energy) of the light received will be much lower than the frequency of emission, owing to the Doppler shift. What has happened to the energy 'lost'? A little consideration shows that it has done work towards the expansion of the universe. The process is equivalent to the well-known adiabatic cooling by expansion, but the mechanical energy, which, in the terrestrial experiment, is transmitted to the walls of the enclosure, becomes in the cosmological application absorbed in the large-scale mass motions of the universe and is hence no longer available (at least in the usual case) for further transformation. This is, of course, precisely the 'sink' for radiation discussed earlier, but we must now consider the significance of the result that the energy of radiation absorbed by the sink is no longer useful.

From a thermodynamic point of view we must therefore distinguish between three types of processes:

(i) Conversion of material energy into radiation.* This occurs principally in stars.

(ii) Conversion of radiative into available material energy. (Heating of interstellar gases and of stars by radiation.)

(iii) Loss of radiative energy in the 'cosmical' sink.

From our observations it is clear that (i) vastly exceeds (ii). Whether (iii) just balances the account is not clear, and different theories give different answers. Kinematic relativity views the problem in a rather different way, but the discussion given applies to most other theories. In any case it is important to realize that here is a curious problem which has possibly not yet received sufficient attention.

The considerations of this chapter are closely connected with a celebrated question of ordinary physics, that of the so-called 'arrow of time'. Although all local physical laws (other than the second law of thermodynamics) are invariant against time reflection, both everyday experience and the second law show that the direction of time is of prime importance. It can then be argued convincingly (Gold, 1958) that the direction of time is in fact determined by the expansion of the universe through its influence on the fate of radiation. Though the connection between the expansion and local phenomena is not always easy to trace, it seems hardly possible that there should be any additional factor determining the arrow of time. This consideration is of prime cosmological importance as it would effectively rule out any model of the universe that did not expand throughout its career.

* Part of this energy appears in the form of neutrinos. Their fate is similar to that of the radiant energy.

CHAPTER IV

THE PROBLEM OF INERTIA

4.1. There are in principle two entirely different methods of ascertaining the speed of rotation of the Earth. A purely terrestrial experiment, such as observing the motion of a Foucault pendulum or of a gyroscope, will measure the rotation of the Earth *dynamically* in that it ascertains the motion of the Earth relative to an idealized inertial frame in which Newton's laws of motion apply. Alternatively, we may survey the sky and ascertain *astronomically* the angular velocity of the Earth with respect to the 'fixed' stars and nebulae. As is well known these two determinations of the Earth's angular velocity yield the same result (at least to the degree of accuracy of the observations). Is this a remarkable fact, a coincidence which demands explanation?

Opinions differ widely on this question. One school of thought holds that the coincidence is an obvious corollary of Newton's laws, or possibly of the theory of relativity, but another school considers it to be a fundamental fact of the first importance, a result the explanation of which must be one of the chief tasks of any really comprehensive theory of dynamics. This second point of view was put forward clearly and most firmly by Mach (1893). Though his outlook on the subject has not been universally accepted, he has greatly influenced many authors, and should his views prove to be right they would be of the utmost importance for cosmology. Accordingly Mach's principle, as it has come to be called, will be discussed here at some length.

If one accepts with Mach that the coincidence of the dynamically and astronomically determined frames of reference is in no way a consequence of Newton's laws (or the theory of relativity), then clearly it becomes necessary to find an explanation of this coincidence, since an accidental agreement of the two frames of reference to the required degree of accuracy is altogether too unlikely. It therefore becomes necessary to postulate a causal connexion between the motion of the stars (and nebulae) and the state of motion of the local inertial frame. In the absence of a third related object the causal connexion must be supposed to act between the two phenomena mentioned. An influence of the local inertial frame on the motion of the stars is not acceptable, and hence it must be assumed that *the local inertial frame is determined by some average of the motion of the distant astronomical objects.* This statement is known as Mach's principle.

4.2. Can we say anything about the nature of the interaction postulated? Since the measure of the inertia of a body is its mass, it seems extremely likely (for reasons of reciprocity) that the effect of the stars should be due to their masses and proportional to them. On the other hand, it is known that the inertia of bodies is unaffected (at least to the accuracy of experiments) by 'local' masses such as the Earth or the sun. Accordingly, it must be assumed that the influence of the distant bodies preponderates. Therefore the inertial influence of matter cannot depend very critically on distance, so that the effect of distant objects will (owing to their much greater number) completely overshadow the effect of near bodies. Hence inertia will not vary appreciably from place to place, simulating an 'absolute' space against which inertia has to be reckoned and making the 'absolute' space a convenient background for local dynamics.

An example of a long-range effect was given in the last section (light of the night sky). If Mach's principle is correct, the inertial influence of matter supplies another connexion between us and the depths of the universe from which, in turn, some information about the universe may be gained. The fact that inertia is the same in all directions suggests strongly that the universe is, on the large scale, isotropic. For if, as an example, the internebular distance in one particular direction were much smaller than in the perpendicular direction, then one might expect a rather different resistance of matter to acceleration in the first direction, exhibiting the anisotropy of the universe as an anisotropy of inertia. (A model of a vector theory of inertia has been proposed by Sciama (1952), but its relation to general relativity is rather unclear.)

4.3. Of even greater importance than this information is a consequence of Mach's principle first discussed by Einstein (1917): 'In a consequential theory of relativity there can be no inertia of matter against space but only inertia of matter against matter. If therefore a body is removed sufficiently far from all other masses of the universe its inertia must be reduced to zero.'

Although Einstein has assumed in this argument that the inertial influence decreases with (and is not independent of) distance, the main point of that argument is valid even without this assumption; Mach's principle implies that not only the motion of the inertial frame, but also the magnitude of the inertia of each body, is determined by the masses of the universe. That this is indeed a necessary consequence follows from a consideration of the motion of a particle in a completely empty universe.

Since there are no masses except itself, its motion must, according to Mach's principle, be completely unaffected by inertia (the self-force being independent of its motion). If we now consider a second particle of much smaller mass than the first one also present, it is inconceivable that the existence of this one particle, however light, should restore inertia to its customary role. Its effect must be slight, and its existence can only lead to a small resistance of the first body to acceleration relatively to the second. This implies that the *magnitude of the inertia of any body is determined by the masses of the universe and by their distribution*. We may express this differently: It is universally agreed that the gravitational effect of a particle is only determined by its own properties, but if Mach's principle is correct then its inertia depends also on the state of the universe. Accordingly, the ratio of the two effects, the constant of gravitation, contains information about the universe. An appreciation of the significance of the value of this constant for the characterization of the universe must wait for a consistent mathematical theory of Mach's principle, but it is clear that if at any time the universe was in a state very different from the present one then Mach's principle implies that the 'constant' of gravitation had a different value at such a time. Similarly, if the universe did not satisfy the cosmological principle, the constant would vary from place to place. Any theory which assumes a varying universe must deal with the possibility that the constant of gravitation varies; or otherwise make the deliberate assumption that the constant is independent of the structure of the universe, thereby denying Mach's principle.

4·4 We have seen that if Mach's principle is correct then deductions of importance to cosmology can be made. What can be said about whether or not it is a correct hypothesis?

A number of objections have been raised against it by various authors. It is widely believed that Mach's principle is of philosophical rather than physical content, and that hence it can be of no physical relevance even if it were logically and philosophically tenable. This view, based partly on Mach's own outlook and his position in philosophy, seems to be incorrect on the interpretation here given. Not only has it been possible to make deductions of physical significance from it, but it should be possible to test its correctness experimentally. Although the universe seems to be isotropic on the average there are local deviations from isotropy (such as our own galaxy). These should show themselves in small anisotropies of inertia, which may be expected to be of order 10^{-7}. This is on the very verge of what is attainable with present-day apparatus. Such experiments as have been made are inconclusive. Far more work will have to be done before certainty can be attained, but it is clear that any hypothesis that is capable of experimental verification (even if only in principle) is of physical and not of philosophical content.

Of greater importance is the alternative view to Mach's, put forward as the 'hypothesis of absolute space'. Newton introduced this concept and defined 'absolute' space by its inertial properties. If we assume such a background against which all dynamics must be considered, then the fact that the stars do not rotate about us is a simple consequence of the laws of motion. What is more difficult to explain on this basis, however, is why there is no *unaccelerated* motion of the celestial bodies simulating a rotation. Such a motion would be exemplified by a state of uniform shear in which all bodies move with uniform velocity parallel to, say, the x-axis, their velocities being proportional to their y-coordinates. A system in this state would not conflict with Newton's laws of motion but would appear to be, more or less, in a state of rotation from any point within it. That the system is not in fact in such a state is a specific property of the universe revealed by astronomical observations, and not a consequence of the laws of dynamics.

The 'absolute' space concept in this crude form belongs of

course to the pre-relativistic era. Special relativity, however, does not lead to any serious changes. There is a triply infinite set of inertial systems connected by the Lorentz transformations. All other frames of reference are not inertial, and the theory does not connect the selection of the inertial systems of reference with the motion of the masses of the universe.

The significance of general relativity in this connexion is rather more difficult to appreciate. The field equations of the theory unify the gravitational and inertial fields, and appear to give expression to Mach's principle, since they imply an influence of massive bodies on the inertia of other bodies. As Einstein (1917) himself was careful to point out, the postulate of the relativity of inertia cannot be satisfied by merely admitting that the masses of bodies *influence* inertia, it demands that they should *entirely cause* it. The field equations by themselves cannot do this (though they may be consistent with such a demand), since the boundary conditions too have to be chosen appropriately. It seems to be difficult to make such a choice of boundary conditions, and hence general relativity has not been reconciled with Mach's principle in this way. Einstein points out that, with the frequently adopted boundary condition that space should be Galilean at infinity, a particle in the absence of all other masses would have almost the same inertia as in the presence of all the bodies of the universe. For this reason he introduced the so-called cosmological constant in the hope of reconciling general relativity with Mach's principle. This hope was, however, not fulfilled (cf. p. 98). It is now generally agreed that general relativity by itself does not incorporate Mach's principle in detail, though it can construct models which, in a certain sense, satisfy Mach's principle 'overall'. The whole range of the models of relativistic cosmology (discussed in detail in Chapter x) share this property which arises in the following way: The masses of the universe are supposed to be moving along certain simple non-intersecting paths, satisfying the cosmological principle. At each point the matter moving through it defines an inertial system. When we examine any local deviation from this idealized state our boundary conditions are that at a large distance the system is in the undisturbed state. Apart from small local effects then, the inertial frame at each point is one in which all distant bodies seem to be moving purely radially and isotropically. In a certain sense Mach's principle is then satisfied at each point, but we are not told how any particular body causes inertia. This indeed would be possible only to a limited extent in a non-linear theory such as general relativity. This theory also gives information on how local inertia is affected by massive bodies superposed on the general distribution. General relativity, together with suitable cosmological assumptions, may therefore contain all there is in Mach's principle, making further research rather pointless. From a somewhat different point of view it may be said against the hypothesis of the relativity of inertia that it has not proved possible to put it directly into a mathematical theory. This point, emphasized by Hoyle in a recent paper (1948), implies that only very few deductions can be made from the hypothesis, and even if it is correct it cannot be used to make predictions in practical cases. Hence there are hardly any experiments which are conceptually sufficiently simple to allow of a qualitative answer to be found without mathematical theory and yet capable of being performed in practice. Even if Mach's principle is correct, other theories are therefore required to deal with experimental and observational evidence.

Other objections have been brought against Mach's principle but they do not seem to be of a very serious nature. It has sometimes been said that the transverse motions of the 'fixed' stars invalidate Mach's reasoning. Not only are these motions so small that they could at most lead to discrepancies of the order of a few seconds of arc per century between the dynamically and the astronomically determined frames of no rotation, leaving the coincidence virtually as striking as before, but in any case it would be wrong to link Mach's arguments to the fixed stars now that it is known that the masses of the universe reside much further away in the external galaxies. These do not possess any observable transverse velocities and they, as the most distant objects observed, now take the place of Mach's 'fixed' stars. In any case, all that is required is that a suitably chosen average of the motions of the celestial objects should agree with the dynamically ascertained inertial frame, and the motions differ so little from each other that the method of averaging hardly affects the result.

Of similarly little significance is the remark that the discovery of the limiting role of the velocity of light (special relativity) removes the necessity for Mach's principle, since any 'true rotation' of the stars about us would correspond to enormous velocities far in excess of the velocity of light for distant stars. This rather begs the question, since special relativity only applies to inertial systems, and the whole problem is why the stars seem to be at rest in an inertial system. The type of transverse velocity which would, according to this remark, arise is in any case of the type to which the arguments of special relativity do not apply. It is, in fact, just the type of motion actually observed from the rotating Earth. A star only 10 light-years distant has a transverse velocity relative to axes fixed in the Earth of 20,000 times the velocity of light and is nevertheless observable. The special theory of relativity does not deny the existence of purely geometrical velocities in excess of the velocity of light. For example, two rulers inclined at a very slight angle to each other, one fixed and the other moving slowly normally to the first, will have their point of intersection move at a velocity which can be made arbitrarily large by making the angle between the rulers sufficiently small. The special theory of relativity cannot therefore be said to make Mach's principle superfluous.

This chapter may then be summed up by saying that the postulate of the relativity of inertia (Mach's principle) is intellectually agreeable in many ways, and seems to some authors to be inescapably true. Others regard it with suspicion, since it has not been possible so far to express it in mathematical form (not even in general relativity), and since it has not so far been verified experimentally. If the postulate is correct then it implies a very strong connexion between the depths of the universe and laboratory physics. A displacement of the observer in space or time resulting in a substantial change in the picture presented to him by the universe would lead to great changes in his local system of dynamics such as (at least) a change of the constant of gravitation.

CHAPTER V

OBSERVATIONS OF DISTANT NEBULAE

5.1. The two preceding chapters have been concerned with information about the universe based on observations that are easy to carry out but extremely difficult to interpret. The interactions discussed were strong but obscure, and many authors doubt, at least in the case of Mach's principle, whether any information about the universe is conveyed at all. The present chapter deals with observations which, it is almost universally believed, are of great importance to cosmology. It is information which can be interpreted without great conceptual difficulty, but can only be gathered by the most refined use of the greatest astronomical apparatus of our time. It has frequently been considered to be the only observational material of importance to cosmology. This view seems to us to be mistaken. Of great significance as these intricate observations are, they are by no means the only information concerning the universe, and in our opinion the data discussed in the other chapters of this part of the book are of similar and comparable importance.

It is clear that of all astronomical observations those concerned with the most distant objects must be of the most immediate significance to cosmology. It is true that observations of our astronomical neighbourhood are necessarily much more accurate and detailed and that, if properly interpreted, they lead to results of great importance to the subject of cosmology. However, the information obtained from the more distant objects is in many ways so surprising and interesting that it has provided a great stimulus to the development of cosmological research. It will be discussed in this chapter, and consideration of the results of modern astrophysical research on the data of astronomical observations of our galaxy will be deferred to the next chapter.

The survey of regions well beyond the borders of our own galaxy may be considered to have begun in the early 1920's. The researches of Slipher, Shapley and Hubble then established that many of the objects of ill-defined boundaries that had long been known under the name of 'nebulae' did not belong to our galactic system but were very far beyond its borders. The establishment of mile-posts and the charting of these vast regions took place during the 1920's and 1930's. The conquest of the 'realm of the nebulae', as Hubble so appropriately calls it, forms one of the most fascinating stories in the development of science. Only a brief survey of the results can be given here.* The most powerful tools for research at present available are the 200-in. Mt Palomar telescope, the 100-in. Mt Wilson telescope and the 48-in. Schmidt telescope on Mt Palomar, together with the equipment of the Lick Observatory.

5.2. The number of measurements that can be carried out on a nebula depends greatly on the amount of light we receive from it. According to this criterion the extra-galactic nebulae may be divided into three groups depending on their apparent magnitudes,† m:

(*a*) $23 > m > 19$. Nebulae in this class are so faint that only their positions and approximate magnitudes can be determined. Nebulae fainter than $m = 21$ cannot be observed usefully on the 100 in. telescope.

(*b*) $19 \geqslant m > 13$. These nebulae emit enough light for a good deal to be known about their shapes, in addition to their positions and magnitudes. It is also possible (though only with great difficulty, at least in the case of the fainter ones) to determine the Doppler shift in the spectrum of these nebulae.

(*c*) $13 \geqslant m$. Fairly complete information is available about the few thousand nebulae in this class. Many of them have been resolved into stars in their outer regions though only a few nuclei have been so resolved. Since the absolute luminosity (rate of emission of light) is known for some types of stars observed with fair accuracy from our astronomical neighbourhood, the distances of some of these nebulae are known with considerable precision.

* For a fuller description reference should be made to the books and papers listed in the bibliography.

† The apparent magnitude m is connected with E, the energy/cm.^{2}sec. received, by the relation $m = \text{const.} - 0{\cdot}4 \log_{10} E$. The value of the constant depends slightly on the spectrum received.

From the observations of nebulae in all three groups the following information may be abstracted:

(i) *Isotropy of distribution*

Nebulae of any magnitude class seem to be distributed around us without preference for any direction. Considerable numbers of nebulae have to be considered, however, before irregularities are smoothed out, mainly owing to the clustering tendency of nebulae (see (ii)). Unfortunately, this result cannot be established over much more than half the sky, since nebulae cannot be observed in or near the plane of our own galaxy owing to the obscuration there.

(ii) *Occurrence of clusters of nebulae*

Nebulae are frequently found in groups which are referred to as clusters. They contain any number of nebulae from a few to a few thousand. The angular separation of nebulae in a cluster is usually, though not always, very much smaller than the average angular separation of two neighbouring nebulae of the same magnitude in the general field. This clustering tendency is very strong. If random samples of numbers of nebulae are taken they are very far from forming an error distribution. For samples of a reasonable size it turns out instead that the logarithms of the numbers of nebulae follow approximately an error distribution. A full statistical analysis of clustering is being carried out (Neyman, Scott and Shane, 1955–8).

(iii) *Shape and structure of nebulae*

In the case of the nearest nebulae and in the case of our own galaxy it can be established that they form more or less plane disk-like structures, i.e. the extension of any nebula in one direction is only about one-tenth of its extension in the two perpendicular directions. Although direct proof of this property is impossible for the more distant nebulae, all the indications tend to show that this flat disk-like shape is an almost universal property of nebulae.

The structure of a nebula in its own plane often shows some sort of spiral character, but a substantial fraction of the nebulae appear to be largely uniform within their boundary ellipses (elliptic nebulae). This second type appears to be more common in clusters than among single nebulae. Only a small fraction (about 3%) of nebulae do not fall into either of these groups, but present an irregular appearance. There is some evidence that the spiral character of the majority of nebulae is due to the distribution of the brightest stars only, the remainder being distributed uniformly.

The central part (nucleus) of the nebulae is usually very dense, but does not contain any very bright stars. It has been suggested that the stellar population of these regions differs markedly in its make-up from the population of our neighbourhood which seems to be similar to that of the outer regions of all nebulae. Other features of the outer regions of our galaxy such as star clusters also seem to be repeated in the outer regions of other nebulae. The nucleus contains only a small fraction of the mass of a nebula, though this has been established with precision only in the case of one or two of the nearest nebulae.

The radius of a nebula is a rather indefinite quantity owing to the gradual fade-out of the nebula near its borders, but it is usually between 10,000 and 50,000 light-years,* with 30,000 light-years as a good average figure. The radius can of course only be measured if the distance of the nebula is known, which requires the methods described later for nebulae in groups (*b*) and (*c*).

(iv) *Luminosities and spectra of nebulae*

A direct measurement of the absolute luminosity can only be carried out if the distance is known by some other means, i.e. in group (*c*). On the very plausible assumption that the nebulae of a cluster are close together, not only in angular separation but also in depth, it is possible to find differences of luminosities from observations of clusters and so to find the frequency distribution of luminosities about a mean, which cannot itself be assigned a value by this method.

Using both these approaches Hubble found that the spread of luminosities was not so very great and that averages taken in different regions agreed well with each other.† Accordingly, the procedure may be reversed and the distance of nebulae too faint to be measurable otherwise may be estimated from their apparent

* 1 light-year $\doteqdot 10^{18}$ cm.; 1 parsec $= 3{\cdot}087 \times 10^{18}$ cm. The light-year, giving time elapsed as well as distance, seems to be the best unit for cosmology.

† Average absolute magnitude of nebulae $M = -19{\cdot}5$ or thereabouts.

brightness. Inaccurate though this method is, it is the only way of obtaining any idea of the distances of nebulae in groups (*a*) and (*b*) and has been of great value.

It is estimated that the radiation of an average nebula is equivalent to the radiation from about 10^{10} stars of solar type. This is composed of contributions received from stars of high-, medium- and low-surface temperatures, and it has been suggested that these radiations combine to give a spectral distribution not very different from that of a black body at 5500–6000°. There is, however, still considerable uncertainty attached to this description of an average nebular spectrum, particularly for the violet and ultra-violet parts of the spectrum. The question is of high importance, and a more accurate knowledge of the spectrum would greatly help the interpretation of other data, as will be described later.

(v) *The mass of a nebula*

Knowledge of the total luminosity of a nebula is little help in working out its mass. For in the first instance the luminosity per unit mass varies greatly between stars of different types (the brightest ones being the most efficient radiators), and secondly, part of the mass of the nebula is in the form of non-luminous matter (gas and dust). Though radio measurements may assist in estimating the mass of neutral atomic hydrogen gas, the quantity of molecular hydrogen cannot be determined by direct means. All mass estimates are therefore inferential and uncertain, though perhaps the greatest weight should be given to those based on differential velocities in clusters interpreted in accordance with the virial theorem, Sinclair Smith (1936). The most probable figures for masses come out in the neighbourhood of a few times 10^{44} g. for the largest nebulae, and appreciably less for the smaller ones.

(vi) *The velocity-distance relation*

This famous relation summarizes the results of a series of measurements of the Doppler shift of absorption lines in the spectrum of nebulae. The difficulty of these observations is very great owing to the diffuseness of the source and the faintness of the light received. Only a considerable refinement of technique, developed principally by Humason, enabled him, Mayall and Sandage to extend the range of observations to nebulae as faint as the 18th and even 19th magnitudes.

The shift of the spectral lines of nebulae is almost always to the red and is found, on the average, to conform well to a relation of the type

$$\log_{10} z = 0{\cdot}2m - X, \tag{5.1}$$

where z is the Doppler shift $\delta\lambda/\lambda$, m is the apparent magnitude of the nebula, and X a constant variously estimated to be between 4·7 and 5. If the Doppler shift is interpreted as a velocity v of recession and m as an indication of the distance r (i.e. the intrinsic luminosity of all nebulae is taken to be constant and the light is supposed to be weakened according to the inverse square law), then the relation may be written

$$V = \frac{r}{T}, \tag{5.2}$$

where T is a constant related to X but depending also on the average intrinsic luminosity of a nebula. On this interpretation the radial velocity of a nebula is proportional to its distance.

The deviations of individual nebulae from the relation given by (5.1) is much smaller for faint than for bright nebulae and may be ascribed to random velocities of a few hundred kilometres per second and to the scatter in absolute luminosities. Since the faintest nebula whose spectrum has been taken shows a red shift of 0·2, which may be interpreted as a radial velocity of 60,000 km./sec., the scatter is very small indeed for the fainter nebulae. Even larger red shifts have been determined by Baum who, using photo-electric measures of the continuum, has found a galaxy with $z = 0{\cdot}4$.

The evaluation of T depends on r, which is far less accurately determined than v. The reciprocal of T is known as Hubble's constant, and Hubble originally found it to be 540 km./sec./mega-parsec, corresponding to $T = 1{\cdot}8 \times 10^9$ years. This value was current for many years until Baade showed in 1952 that the distances had been seriously underestimated. Sandage (1958) gives a value of $T = 1{\cdot}3 \times 10^{10}$ years, i.e. a Hubble constant of 75 km./sec./megaparsec.

Modern measurements also tend to show that X is not constant but varies slowly with z. If it is represented by a linear function, the coefficient of z might be around unity, but this figure is very uncertain, partly owing to selection effects (Scott, 1957), since at large distances only exceptionally bright galaxies will be seen.

The significance of Hubble's constant becomes very clear if we assume the motion of the nebulae to be unaccelerated. For on this basis all the nebulae were packed into a small region $1 \cdot 3 \times 10^{10}$ years ago and moved away as though an explosion had taken place there, each with its own individual velocity, the 'fastest' ones covering the 'greatest distance'. That such a picture can be drawn without treating any nebula in a preferential way is a consequence of the special theory of relativity and will be discussed in Chapter x.

The velocity-distance relation has been established both by observations of single (field) nebulae and by observations of members of clusters. In the second method the Doppler shift of the fifth brightest nebula of a cluster was measured so as to avoid the need for averaging and yet to obtain slightly smoother fitting figures than would have been obtained with the brightest nebulae of each cluster. The uncertainties that arise in the constants of the velocity-distance relation are largely due to the fact that for near nebulae the random velocities are of the same order of magnitude as the general velocity (10^7 cm./sec.), whereas for distant nebulae the measurements are far more difficult to carry out, are few in number and contain larger errors (especially the magnitude estimate). Though X is quite well known and lies between $4 \cdot 7$ and 5, the uncertainties in the distance scale affect the estimate of T. The lack of correlation between direction of measurement and deviations from the relations greatly supports the assumption of isotropy, and the form of the red shift-magnitude relation is in striking agreement with the assumption of homogeneity which, as we shall see, permits only motions of this type. In fact it may be said that the velocity-distance relation is the strongest evidence in favour of the cosmological principle. Any theory adopting the cosmological principle immediately arrives at the velocity-distance relation, whereas any theory attempting to do without it must make special *ad hoc* assumptions to explain it.

(vii) *The distribution of nebulae in depth*

This is a question of great observational and interpretational difficulty, but owing to its bearing on the homogeneity postulate is one of great theoretical significance.

The number of nebulae in group (*c*) is too small to permit any inferences of great certainty to be made, though the observations definitely do not disagree with the assumption of a homogeneous distribution. As soon as we leave group (*c*) the apparent magnitude m of a nebula is the only indication of its distance. Consider now a distribution of nebulae in a Euclidean space, the 'nebular density' (number of nebulae per unit volume) being on the average constant. Assume also that the intensity of light is weakened only owing to the inverse square law and that the intrinsic luminosity of all nebulae is the same. Then the number of nebulae at distance less than r varies like r^3, whereas the intensity of light received from nebulae at distance r varies like r^{-2}. Translating this into astronomical terms we have

$$\log_{10} N(m) = 0{\cdot}6m + \text{const.}, \tag{5.3}$$

where $N(m)$ is the number of nebulae brighter than apparent magnitude m.

Comparing this formula with the observations it is found that the fit is very good up to about magnitude 18, but appreciable deviations occur for nebulae fainter than this. It is a primary task of any cosmological theory to yield a relation between $N(m)$ and m which applies over the whole range. The observations consist of taking nebular counts on photographic plates exposed for a certain length of time. The faintest magnitude producing an image after such an exposure has then to be evaluated by rather intricate means, and the results of a number of similar exposures are extrapolated to the number of nebulae per square degree (or over the whole sky) assuming the distribution to be isotropic.

Note that the observations, given in the table below, were all made before 1938.

Table I

m	12·78	18·20	18·47	19·00	19·40	20·00	21·03
$\log_{10} N(m)$	3·233	6·328	6·502	6·777	6·955	7·301	7·777
$F(m)$	4·435	4·592	4·580	4·624	4·685	4·699	4·841
$u\,10^{-5} = 10^{0\cdot 2m-5}$	0·0036	0·044	0·049	0·063	0·076	0·100	0·161
$v = \dfrac{E}{3} F(m)$	3·404	3·524	3·515	3·549	3·595	3·606	3·715

The numbers in the last two rows will only be required in Chapter x. The quantity $F(m) = 0{\cdot}6m - \log_{10} N(m)$ is a measure of the deviation of the actual from the 'common-sense' distribution which, as was shown above, should make it a constant. The constancy of F for $m < 18$ is a strong indication of the correctness of the assumption of homogeneity and greatly supports the hypothesis of the cosmological principle. The variation of $F(m)$ for $m > 18$ which at first sight looks so puzzling is by no means surprising. Four different reasons may be given for this variation:

(1) Although the velocity-distance law has only been established for $m < 19$, any plausible extrapolation to the range $19 \leqslant m < 21$ shows that the red shift is bound to be very considerable in this range (z may run up to 0·25). Such red shifts correspond to a considerable loss of energy which, in addition, is overemphasized by the photographic plate owing to the shift of the spectrum. Light is therefore weakened more than would correspond to the operation of the inverse square law, contrary to the assumption which leads to $F = \text{const.}$

(2) Space may not be Euclidean over such vast distances as correspond to nebulae of $21 > m > 18$ (up to 5×10^8 light-years).

(3) There may be absorbing material in space which reduces and reddens the light that reaches us.

(4) The number of nebulae per unit volume and the average luminosity of nebulae may have been different at the time the light was emitted. In whatever way 'time' may be defined it is clear that the emission took place very long ago.

That reason (1) is operative few if any would deny, and (4) is probably not very important. Different theories disagree greatly on (2) and (3). If the variation of F due to (1) and (4) could be determined, then the residual would contain information about (2) and (3) and might hence decide between the various theories. Unfortunately, the measurement of the Doppler shift becomes impossible, at the present time, just at magnitude 19 where the variation of F becomes interesting. The only way of progress then seems to be to require each theory to formulate the variation of $N(m)$ and hence of $F(m)$ with m, due allowance having been made for the weakening of light by the Doppler shift $z(m)$. The variation of z with m must therefore also be predicted by the theory. A further difficulty arises because whereas a theory may predict the Doppler shift and the loss of energy consequent upon it, there is also an additional correction to be made because it is not the total energy of the radiation that is measured, but its effect on a photographic plate after traversing the atmosphere and the telescope. The photographic efficiency of radiation of a certain total power greatly depends on its spectrum, since red light hardly affects the plate and ultra-violet is absorbed by the atmosphere. The difference between bolometric magnitude (total energy) and photographic magnitude is called the K-term. Since the Doppler shift moves the entire spectrum the K-term will be a function of z, but the nature of this function depends critically on the nebular spectrum. The question is discussed in an appendix to this chapter, but it will be seen that the evaluation of the K-term greatly adds to the difficulties of interpreting the number counts.

The observations of near nebulae serve to calibrate the distance scale, and it is found that the average internebular distance is about 2 to 8 million light-years, i.e. about 2 to 8×10^{24} cm.

(viii) *The Stebbins–Whitford effect*

In 1948 Stebbins and Whitford began measurements of the continuous spectra of distant galaxies by measuring the intensity of light received in a number of bands of wave-lengths. They then claimed to have evidence that the light of distant galaxies was considerably redder than the light of near galaxies when allowance had been made for the reddening associated with the red shift of the spectral lines. One possible explanation of such an effect is that light is reddened in transit by absorption, but this implies an improbably high density of intergalactic matter. The only alternative is that the distant galaxies were much redder when they emitted the rays now received here than near galaxies are at present. Therefore this effect was taken by Gamow and others to be direct evidence for the evolution of the universe and thus against the steady-state theory, but the very size of the effect (appreciable changes in colour over periods of a few hundred million years) was considered surprising. More specific criticisms of the results were given by De Vaucouleurs (1953), and by Bondi, Gold and Sciama (1954). Further measurements by Whitford and

Code (1956) proved conclusively that the original results had been due to the peculiarities of the standard chosen and that there was no evidence for any such change of colour with distance.

Although the existence of the phenomenon out to ranges of a few hundred million light-years has thus been disproved, the Stebbins-Whitford effect is of importance in pointing the way to a type of measurement of potentially great significance to cosmology.

(ix) *Radio astronomy*

Radio astronomy was first linked with cosmology when Gold (1951), opposed by Ryle (1951), suggested that the discrete sources ('radio stars') of radiation in the accessible band (below 5 *m* wavelength) were extragalactic and (except for a few sources in the galactic plane) this is now thought to be so (Ryle, 1958). In 1953 Baade and Minkowski identified a prominent radio source with a faint optical object that was interpreted as two galaxies in collision. This, and the few more recent identifications, show that while most galaxies radiate only weakly in radio wavelengths a few rare ones radiate powerfully. Presumably many of the radio sources observed are such rare powerful emitters, and are thus well beyond optical range; this is made plausible by the shape of the radio spectrum, which is so flat that the red shift has no critical effect on the intensity.

More recently Ryle and his colleagues (1955) used the Cambridge radio interferometer to determine the position and strength of nearly 2000 radio sources at a wavelength of 3·7 *m*. It turned out that away from the galactic plane the sources were isotropically distributed, supporting the view that they were extragalactic. Furthermore, it emerged that the increase of number with faintness was far more rapid than would correspond to a spatially uniform distribution.

Ryle and his colleagues (1955) interpreted their results as indicating an evolution of the universe, radio galaxies having been far more common in the past than at present, and thus disproving the steady-state theory. However the survey of Mills and Slee (1957) in Australia flatly contradicted the Cambridge results. A source-by-source comparison of the two sets of results in the overlapping region, tests by Lovell at Jodrell Bank and a new Cambridge survey at 1·9*m* (Edge, Scheuer and Shakeshaft, 1958) all indicate that the first Cambridge survey was quite unreliable owing to instrumental errors, and that no valid conclusions could be drawn from it. It may, however, be possible with present equipment to determine whether the distribution of radio sources is uniform in depth out to very great distances or whether there is evidence of evolution. These studies will be especially significant if further work such as that of Lilley *et al.* (1957) confirms the estimate of the distances of the sources.

(x) *The mean densities of matter and radiation in space*

In his estimate of the density of matter in space the astronomer is in the first instance guided by his direct observations of luminous matter, i.e. of stars. Since the relation between the mass and the luminosity of stars of each spectral type is well known, and since the frequency distribution of stars of each spectral class is also fairly well known, the stellar matter contained in each nebula may be estimated fairly accurately. Combining this result with the estimate of the mean intergalactic distance previously mentioned, Shapley (1933) obtained an estimate of a few times 10^{-30} g./cm.^{3} for the mean (smoothed out) density of *luminous* (*i.e. stellar*) *matter* in space. This has frequently been misunderstood to imply that the mean density of *all matter* is given by this figure.

A more fallacious view could hardly be imagined, and yet it has been held persistently for many years by a number of eminent authorities. Clearly 10^{-30} g./cm.^{3} can only be a lower limit and nothing else. An upper limit can be obtained by considering what densities of interstellar and intergalactic matter would have produced observable effects which have not in fact been observed. Now there is good reason to believe that the bulk of the non-luminous matter is in the form of hydrogen, and this would not produce any observable absorption lines, since most of the hydrogen would be in the ground state and the Lyman series is blotted out by the atmospheric absorption of ultra-violet light. Considerations of this type can be adduced to show that even a suitably distributed density of 10^{-25} g./cm.^{3} would not be incompatible with observation.

Radio observations show that galaxies are surrounded by haloes whose large volume, in spite of their low densities, implies that they contain quite an appreciable mass. The gentle way in which these haloes shade off into intergalactic space again suggests that the density there may not be many orders of magnitude lower than galactic densities. Since the volume of intergalactic space exceeds the volume of the nebulae by a factor of the order of one million, intergalactic matter may make a large contribution to the mean density. Altogether it seems at present most likely that the mean density of matter in space is within a factor of 5 from 10^{-29} g./cm.^{3}.

The mean density of radiation may be estimated from the light we receive from astronomical objects outside our own galaxy. We must exclude the light from members of our own galaxy, since the average distance of a point of space from its nearest galaxy is of the order of a third of the mean intergalactic distance, whereas we are well within our own galaxy. The extra light due to our privileged position must be disregarded before we can obtain average values. It should also be noticed that for an observer well outside any galaxy the contribution of the nearest galaxy does not form a substantial fraction of the light received. For, as was discussed in Chapter III, the contribution of the more distant nebulae will be of great importance owing to their large numbers.

An estimate of the mean density of radiation, carried out in this way, leads to a value of about 3×10^{-13} erg/cm.^{3}, corresponding, according to the theory of relativity, to a mass density of radiation of about 3×10^{-34} g./cm.^{3}. This is only a minute fraction of the mass density of matter.

An independent method of arriving at this figure consists in taking Shapley's value of the mean density of luminous matter in space and multiplying it by the average rate of emission of light per unit mass which is known from our own galaxy. In this way we arrive at a rate of emission of 10^{-30} erg/cm.^{3}sec., averaged throughout space. The light will travel through space until it is absorbed or until its energy is reduced substantially by the Doppler shift due to the operation of the velocity distance relation. The time constant for the second process is $T = 4 \times 10^{17}$ sec., and the first process is probably less important. Accordingly, the light will effectively contribute to the mean density of radiation for a period T, which gives a value of 5×10^{-13} erg/cm.^{3}, in close agreement with the value given above.

(xi) *The rotation of the nebulae*

The most immediate explanation of the disk-like shape of the nebulae is that they are in a state of rotation sufficiently rapid to prevent further gravitational contraction in the plane of the disk. It is difficult to resist the impression that the spiral arms are connected with a state of rotation, and this is assumed in all the theories of them. The rotation of our own galaxy can actually be observed as part explanation of the proper motions of the stars. Finally, the rotations of some of the nearest nebulae have been measured, although the observations are very close to the limits of present possibilities of measurement. It seems safe to suppose that the nebulae are rotating, and the indications are that the period of a rotation is usually of the order of 10^8 years.

5·3. This concludes our highly condensed account of the results of one of the most fascinating and fruitful series of observations in modern science. We have yet to discuss what bearing these results have on cosmology. The subject of our study is the whole of the theoretically observable universe. The observations of a region of radius a few hundred million light-years around us are certainly of interest, just as any observations whatever must be of interest to a science so comprehensive as cosmology. But there is the possibility that these observations may be far more significant, the possibility that they concern a region large enough to be a fair sample of what the universe is like anywhere. This question is clearly of fundamental importance to our subject.

The cosmological principle postulates the homogeneity of the universe in the sense that any two sufficiently large parts of it are alike in all general physical properties, but does not say how big a region must be before it is 'sufficiently large'. Observations show that regions of radius 10,000 light-years are certainly not sufficiently large, since some such regions contain a nebula and are then very different from others which do not. But what are the indications that the regions covered by the observations, 5×10^9 light-years across, are 'sufficiently large' within the meaning of the postulate? Hubble discusses this question in detail, and his arguments in favour of the 'fair-sample hypothesis' are very powerful indeed. It is true that the question cannot be decided for certain until the entire universe has been explored, but science requires, in this as in many other cases, a working hypothesis, the formulation of which cannot wait until certainty is achieved. There can be little doubt that the 'fair-sample hypothesis' is the most direct and fertile one that can be inferred from the observations. The great homogeneity of the region observed and the total absence of any indication of a boundary (other than that given by the velocity-distance relation, which in turn is not only compatible with, but actually demanded by, the cosmological principle) strongly point towards the correctness of the hypothesis. The magnitude of the red shift ($\delta\lambda/\lambda$ about $\frac{1}{4}$ to $\frac{1}{2}$) for the most distant objects observed is in itself suggestive that we have surveyed a substantial part of all that is accessible with any ease, and unless a 'substantial part' is 'sufficiently large' the cosmological principle is devoid of meaning for lack of comparable components of the universe.

We may therefore regard it as a justifiable extrapolation from the observed facts to say that the observations cover a fair sample of the entire universe, indicating that it has the following properties:

(i) It contains matter of mean density about 3×10^{-29} g./cm.^{3} which is largely (or partly?) condensed into nebulae, about 3×10^{24} cm. (3×10^6 light-years) apart.

(ii) The nebulae are mainly disk-shaped, of radius about 3×10^{22} cm. (3×10^4 light years) and thickness about one-tenth of the radius.

(iii) The universe is in a state of expansion in the sense that the nebulae are moving away from each other, the relative velocity of any two nebulae being their distance divided by the constant T which is about 4×10^{17} sec. ($1{\cdot}3 \times 10^{10}$ years).

(iv) In addition to this general motion the nebulae have random proper velocities of order 10^7 cm./sec.

(v) The mean density of radiation is about 3×10^{-13} erg/cm.^{3}, so that its mean mass density is about 3×10^{-34} g./cm.^{3}.

(vi) The nebulae show a strong tendency to cluster but are otherwise randomly distributed.

(vii) The nebulae show great similarity amongst themselves. They are probably all rotating and many show a spiral structure.

APPENDIX

The K-term of the red shift

The atmosphere of the Earth and the telescope scatter and absorb light selectively, i.e. some wave-lengths (notably ultra-violet) are severely affected, whereas others (mainly the longer wave-length) are almost unimpeded. The photographic plate is even more highly selective and is darkened only by a certain range of wave-lengths, which, for the most sensitive plates, are usually confined to the blue and green parts of the spectrum. The combined effects of the characteristics of the atmosphere, the telescope, and the plate are therefore that the darkening of the plate due to a nebula depends not on the *total* energy density of its radiation in the neighbourhood of the Earth, but only on the energy density in the green and blue. The efficiency of the observing methods hence depends very sensitively on the nebular spectrum. Now quite apart from the fact that the spectra of different nebulae are somewhat different from each other, the spectra of the more distant nebulae are systematically shifted towards the red owing to their Doppler shifts. The individual differences may be dealt with by averaging, but the systematic differences are of a more serious nature, since they affect our estimates of the distances of the nebulae. A good deal of work has been done on the correction which has to be applied to the photographically observed magnitudes in order to obtain bolometric magnitudes by allowing for the alteration of the spectrum due to the Doppler shift z. This correction term is known as the K-term. The chief difficulty in evaluating it is our lack of knowledge concerning the average nebular spectrum. It is true that the visible part of the spectrum of near-by nebulae has been measured with fair accuracy. The ultra-violet light, however, is scattered by the Earth's atmosphere and hence cannot be measured. This is most unfortunate since the ultra-violet part of the spectrum is shifted, in the case of the more distant nebulae, to just that part of the spectrum to which the photographic plate is most sensitive. (With modern photographic emulsions, which are far more sensitive in the red, the difficulty is minimised. These were, however, not available before 1940, and all the number counts tabulated on p. 41

were taken with blue-sensitive plates.) It has generally been assumed that the spectra of nebulae are similar to that of a black body at between 5200 and 6000° K. This assumption, which is often stated as a similarity between the nebular and solar spectra, is pretty well established as far as the visible part of the spectrum is concerned, and is very likely correct for the infra-red. We have, however, definite evidence that the solar radiation in the ultra-violet is very much stronger than that corresponding to a black body of 6000° K., and may well exceed that of a black body of 7000° K. This difficulty makes any estimate of the K-term very uncertain.

An additional uncertainty is introduced through the variations in the nebular spectra. As said before these may be averaged out, but the weighting of the different types of spectra will be different according to whether they are due to near nebulae (ultra-violet ineffective) or distant nebulae (chief contribution due to the blue, which was emitted as ultra-violet).

The table below has been computed by Hubble (1936) on the assumption of a black-body spectrum, but considering different temperatures of the black body.

z	6500° K.	6000° K.	5500° K.
0·05	0·08	0·09	0·12
0·10	0·15	0·18	0·25
0·15	0·24	0·28	0·38
0·20	0·33	0·40	0·52
0·25	0·44	0·53	0·68
a	0·68	0·85	1·14
b	0·66	0·85	1·07
$1+\frac{1}{2}a^2-b$	0·54	0·51	0·58

The first five rows represent Hubble's results, while the next two contain the coefficients a, b of the expansion

$$K = 5 \log_{10} (1 + az + bz^2 + \ldots),$$

and the last gives a combination of these two coefficients which will be found to be of importance in certain applications (p. 113).

CHAPTER VI

ASTROPHYSICAL AND GEOPHYSICAL DATA

6.1. The researches, some of the results of which will be described here, relate to stars in our own galaxy, mostly in the astronomical neighbourhood of the sun, to the solar system and to the Earth itself. The two main questions they are concerned with are the age and the chemical composition of these objects. On the plausible assumption that the material of the galaxy in our neighbourhood is typical of the accumulation of matter anywhere in the universe, its chemical composition will be a good guide to the chemical composition of all the matter of the universe, and is therefore of interest to cosmology. The ages of the near astronomical objects and of the Earth turn out to be of even greater relevance and are amongst the most crucial data for cosmology. It is not immediately obvious why these data should be so important, and the reason for this is as follows: Since it cannot be assumed that the Earth and the stars can be of earlier origin than the universe (especially if the origin of the latter was catastrophic) their ages supply *a lower bound* to the 'age of the universe'. Now any theory that contemplates a definite origin of the universe must assign a present age to the universe which is in some way related to the time constant T of the velocity-distance law. In fact it turns out that it is not easy to make the universe much older than this constant T, which according to present observations is about $1{\cdot}8 \times 10^9$ years. If the lower bound obtained from a study of the Earth and the stars were less than T, no difficulty would arise, but it appears that the lower bound exceeds T, possibly by a substantial margin. Hence the great importance of these measurements and considerations.

A considerable variety of methods is available for determining the ages of the Earth and other astronomical objects. It is desirable to emphasize in each case the precise significance of the age determined, together with the nature of the time-keeping mechanism employed.

(i) *The age of terrestrial rocks determined by means of the abundances of radioactive materials*

Uranium and thorium are very long-lived radioactive materials. Their decay constants are well known, and so are the end-products (identifiable lead isotopes) of the radioactive chains of which they are the heads. If a uranium or thorium compound is found in a piece of rock, together with its decay products, the mass ratio of end-products to the uranium (or thorium) is a unique function of the time since the materials have been locked together, unable to separate. Hence this is a determination of the time which has elapsed since the rock has been solid.

The oldest known samples lead to ages (on the nuclear time-scale) of about $2{\cdot}1 \times 10^9$ years. Statistical evaluations of isotope ratios are a matter of some controversy, leading to ages between $2{\cdot}5$ and 4×10^9 years. According to all theories of the origin of the solar system the existence of the Earth must have antedated the formation of solid rock, so that it seems very likely that the age of the Earth is at least $2{\cdot}6 \times 10^9$ years, and probably rather more.

(ii) *The age of meteorites determined by means of the abundances of radioactive materials*

Measurements of the same character have been carried out on meteorites. In particular, the helium content (helium is a by-product of the radioactive decay) has been compared with the uranium and thorium contents. The first measurements resulted in very large ages, but then it was suggested that some helium might be due to cosmic ray bombardment. The He^3/He^4 ratio, which determines the proportion of He due to cosmic rays, has been measured for several meteorites. When this is allowed for, the ages turn out to be a few times 10^9 years.

(iii) *The evolution of the stars*

The theory of stellar structure and evolution has made a great deal of progress in recent years and can now be used to give results of interest to cosmology.

It is now well established that the main source of energy in the stars is the thermonuclear conversion of hydrogen into helium.

Through a self-adjusting temperature mechanism the generation of energy takes place in the deep interior at the same rate as the star loses energy from its surface by radiation (the luminosity of the star). The mechanism of energy transport in the star ensures that the luminosity is a definite steep function of the mass (varying between the third and fifth power of the mass), and also varies slightly with the radius of the star. Owing to the steep dependence of the luminosity on the mass, the rate of generation of energy per unit mass is far greater in massive than in small stars. Accordingly, the time interval during which a star converts an appreciable fraction of its hydrogen into helium is very much shorter for the large than for the small stars.

The sequence of events which follows when a good deal of the hydrogen of a star has been turned into helium is not very clearly understood, but it seems certain that, frequently if not always, the radius of the star would increase very greatly.

Observations show that the great majority of the stars in our galaxy obey the relations between the three observable quantities (luminosity, radius and mass) predicted by theory for uniform stars with large hydrogen but small helium contents. Among the more massive stars, however, a sizable minority have a very much greater radius, but no such distended stars of medium or small masses are known.

It seems, therefore, very reasonable to explain these facts by assuming that only some of the massive stars have had time to convert a sufficient proportion of their hydrogen into helium for the extension of radius to occur, whereas none of the smaller stars have had time. Accordingly, the minimum mass associated with the extended radii determines the age of the oldest stars in our part of our galaxy. More precisely, 'age' means here the lapse of time, since these stars become sufficiently condensed for the conversion of hydrogen into helium to supply the bulk of the energy lost by radiation. This period of time turns out to be between 3×10^9 and 8×10^9 years. It is clear, however, that many of the massive stars have either condensed more recently or have become so massive only more recently through the accretion of interstellar matter.

Although the time concerned has been measured largely by the thermonuclear process of energy generation it also depends on the mass-luminosity relation and hence on atomic and gravitational phenomena.

(iv) *Stellar dynamics*

Although many attempts have been made to use the characteristics of stellar motion to obtain age estimates, no very certain results have emerged. Jeans' pioneer attempts led to ages of 10^{12} to 10^{13} years and have been proved wrong. Modern work has concentrated mainly on the question of the dissolution of star clusters through the tidal action of the galaxy and through the 'accidental' attainment of the velocity of escape by individual members by gravitational interaction. The value of this work is somewhat impaired by its neglect of the dynamical action of interstellar matter. Also, since the mechanism of the formation and the early state of such clusters are very ill understood, it is doubtful whether the investigation of one or two important but possibly not predominating forces tending to dissolve the cluster can lead to a reliable estimate of its age. The estimates for times of dissolution so obtained range from 2×10^9 years upwards, and the minimum period of 2×10^9 years is probably significant.

Other dynamical arguments concern the small eccentricity and the frequency distribution of the separation of binaries. Jeans' original arguments have been shown to be fallacious, and more modern investigations do not lead to any figures of cosmological interest.

Similarly, the velocities of the stars relative to the average of their neighbourhoods seems to be related to the masses of the stars in a rough way so as to lead to an approximate equipartition of energy. This 'thermodynamic state', it has been argued, supports a long time scale. However, it seems much more plausible that the stars with small velocities have accreted much interstellar material (the accretion rate varies like the inverse cube of the relative velocity of star and gas) and have hence become massive.

Altogether, it does not seem that stellar dynamics can at present throw any very reliable light on the age of the galaxy.

(v) *Interstellar dust*

Interstellar material, although largely gaseous, seems to contain a great deal of dust (small solid particles). Some work has been done on the origin of this dust, and in particular on its possible condensation from gaseous matter.

From the calculated rates it appears that there is no great difficulty in arriving at a plausible dust/gas ratio in a period of 10^9 to 10^{10} years, but both observation and theory are too vague to lead to any better indication of the age of the galaxy.

The result of all these observations and considerations may be summed up in the statement that there are various methods to show that the age of our galaxy is between 3 and 15×10^9 years, and the age of the Earth is between 2·6 and $4{\cdot}5 \times 10^9$ years.

Comparison with the nebular data indicates therefore that the age of our astronomical neighbourhood is at least $1{\cdot}5T$ and possibly $2T$ or $3T$, using Hubble's value of T (p. 39). It is immediately evident that these ages are then of great significance for cosmological theories, although they refer only to our neighbourhood, since the age of the universe (especially if its origin was catastrophic) cannot be less than the age of any part of it, however small. If however modern values of T are used, then none of these ages exceeds $\frac{1}{2}T$ except for the age of our galaxy which at most equals T. Though these ages are of cosmological interest, as they are of order T, they now have none of their former significance. They do, however, rule out theories in which the age of the universe is markedly less than T.

6.2. The chemical composition of stellar and interstellar material forms one of the most important problems of astrophysics. There are three methods for determining the composition of stellar material, viz. spectroscopic examination of the light emitted by the surface, an exact determination of the opacity of the stellar material by means of the mass-luminosity-radius relation, and by precise evaluation of the process of energy generation which depends on the chemical composition. In all three cases small traces of medium and heavy elements affect the result very critically, whereas the hydrogen abundance is not quite so directly important.

It was therefore for a long time greatly underestimated, and only recent researches have definitely shown that hydrogen predominates very much. There is now every reason to believe that hydrogen and helium together account for well over 90 % of the mass of a star and may account for over 99 %. Carbon, nitrogen, oxygen and the metals form most of the remainder. Their existence is, in spite of their tenuousness, of the greatest importance. The thermonuclear generation of energy depends in large stars on the presence of carbon and nitrogen acting as catalysts; the opacity in the outer regions of a star depends greatly on the metals, and their absorption lines are most prominent in the stellar spectra. The helium/hydrogen ratio is much more variable and depends on the size and age of a star, since the process of energy generation converts hydrogen into helium.

The composition of the interstellar material can be investigated spectroscopically, and it, too, seems to consist almost entirely of hydrogen. This conclusion is confirmed by the theory of stellar evolution in which the accretion of interstellar matter plays a large part. The observations of the stars indicate very strongly that this accreted material is largely hydrogen.

If these results are characteristic of the composition of matter everywhere, then the universe consists mainly of hydrogen, the simplest of all elements. This is very satisfactory from the point of view of the simplicity postulate.

It must be admitted, however, that there is no compelling reason for assuming that our astronomical neighbourhood is typical of the entire universe. The similarity of stars in other galaxies to near stars is favourable to this hypothesis, but the evidence on this point is not too strong. Our ignorance of the composition and density of intergalactic matter is almost complete, and, even if its density were very low, it might nevertheless contain the bulk of the matter of the universe owing to the great size of the intergalactic regions. However, it seems most probable that, if the intergalactic regions contribute substantially to the amount of matter in the universe, their contribution should consist almost entirely of hydrogen, the most inconspicuous of all elements. The predominance of hydrogen may therefore conceivably be even stronger than local observations suggest.

The predominance of hydrogen and the fact that its nucleus, the proton, is the simplest possible one have led for many years to speculations that the nuclei of all the other elements might have been built up from hydrogen nuclei by processes that can be traced out in detail. To put it differently, while the question of the origin of hydrogen might remain a problem outside physics, the origin of all other elements, it is thought, could be shown to result from the operation of ordinary physical laws.

The problem of nucleogenesis, as it is called, divides naturally into two parts:

(i) What are the conditions in which hydrogen can be transmuted into other elements?

(ii) When and where are these conditions realized?

As far as the building up of helium is concerned the answers to these questions have long been known. The most usual stars, the main sequence stars, derive their energies from the thermonuclear fusion of hydrogen into helium at temperatures of 10^7 to 3×10^7 degrees. For many years, however, the nuclear physicists thought that the building up of elements beyond helium required very much higher temperatures, of 10^9 degrees or more, at very high densities (10^{10} g./cm.^{3}).

Two tentative answers were then given to the question of where such conditions might exist. The very peculiar stars known as supernovae, which suddenly become extremely bright (about 10^{10} times as bright as the sun) and fairly rapidly dim again, occur on an average once every 200 years or so in each galaxy. There is a plausible theory of these stars which suggests that they are explosions from a stellar condition such as is required for nucleogenesis, and in this explosion scatter the newly formed elements through interstellar space. However, with any reasonable estimate of the mass transmuted in each such explosion, the total amount of elements formed in this manner turns out to fall short of the abundance required for the nuclei between helium and iron. For the much scarcer very heavy nuclei the amount formed is likely to be sufficient.

The second location then suggested followed from the consideration that if at present there were no suitable furnaces in the universe they may have existed in the past. Hence, in agreement with various cosmological models, it was thought that the universe had a highly condensed very hot early stage in which the nuclei were formed. This theory was in particular discussed by Gamow, Alpher and Herman. The supposed absence of presently existing hot dense places led thus to the conclusion of great cosmological significance that the heavy elements were a relic of a suitable early stage of the universe.

In recent years, however, Cameron drew attention to the fact that neutrons would appear in stellar nuclei at a certain stage of stellar evolution (red giant stars) and that they would rapidly transmute light elements into heavy ones. G. and M. Burbidge, Fowler and Hoyle have then been able to account with remarkable accuracy for the observed abundances of nuclear species, using both the relatively common red giants and the rare supernovae as the location of nuclear transmutation.

In the course of constructing this theory, Hoyle was forced to postulate the exact value of a previously ill-determined nuclear energy level in ^{12}C which has since been verified experimentally.

Since it has also been shown that any hot dense early stage of the universe could not have left us any nuclei heavier than helium, the origin of such nuclei is no longer a question of cosmology.

It might however be said that the abundance of helium may conceivably be greater than would be accounted for by ordinary stellar transmutation and so might have to be explained on a cosmological basis, but the evidence as yet is far too slight to merit serious consideration now.

CHAPTER VII

MICROPHYSICS AND COSMOLOGY

7.1. Present-day physics has been reasonably successful in showing that the phenomena observed in everyday life can be explained satisfactorily in terms of the interaction of 'elementary' particles (protons, neutrons, electrons, etc.). It is true that in many cases it is beyond the present power of mathematics to show that the combined action of these particles would actually lead to the observed behaviour. But since no evidence to the contrary has been forthcoming it is universally assumed that, say, the complicated behaviour of liquids is due to the complexity of the interaction of elementary particles of simple properties rather than to some unknown agency. It follows that one cannot expect to find phenomena throwing light on cosmology in the physics of objects of a size comparable to, say, 1 cm., but may find such phenomena in the realm of microphysics.

Unfortunately present-day understanding of atomic and nuclear physics is not deep enough to indicate in any direct way which results are of particular significance for cosmology. There are, however, a few numerical 'coincidences' arrived at by combining cosmical, 'ordinary size' and atomic measurements. These coincidences are very striking and few would deny their possible deep significance, but the precise nature of the connexion they indicate is not understood and is very mysterious. Some cosmological theories (see Chapter XIII) attempt to base themselves on these coincidences, but none of these is widely accepted.

The coincidences may be constructed as follows: If as usual,

e = electronic charge,

m_e = mass of electron,

m_p = mass of proton,

γ = constant of gravitation,

c = velocity of light,

ρ_0 = mean density of matter in the universe,

T = reciprocal of Hubble's constant,

then the following pure numbers can be formed:

$$\frac{e^2}{\gamma m_p m_e} = 0{\cdot}23 \times 10^{40}, \tag{7.1}$$

$$\frac{cT}{\left(\dfrac{e^2}{m_e c^2}\right)} \doteqdot 4 \times 10^{40}, \tag{7.2}$$

$$\frac{\rho_0 c^3 T^3}{m_p} \doteqdot 10^{80} = [10^{40}]^2, \tag{7.3}$$

$$\gamma \rho_0 T^2 \doteqdot 1. \tag{7.4}$$

Expression (7.1) gives the ratio of the electrical to the gravitational force between an electron and a proton. The value is well established and should not be in error by more than 1 %. In expression (7.2) the numerator is a length derived from Hubble's constant and may be termed the characteristic length of the universe, but is known more commonly, though less exactly, as the radius of the universe. The denominator is a length which is known as the classical radius of the electron, but may be interpreted as the range of nuclear forces. The numerator in (7.3) has the dimensions of mass, and may be called (rather loosely) the mass of the universe, since it is the product of its mean density and the cube of its 'radius'. The denominator being the mass of a proton (or neutron), (7.3) may be termed the number of heavy elementary particles in the universe. Expressions (7.2) and (7.3) are not as well determined as (7.1), but (7.2) should not be in error by more than a factor of 2, and (7.3) by a factor of 50. Equation (7.4) may be in error by a similar factor and can be interpreted in the following way: The gravitational potential energy of the rest of the universe in the field of a single particle of mass m is roughly equal to the expression

$$\gamma m \text{ ('mass of the universe')/('radius of the universe')} = \gamma m \rho_0 c^2 T^2.$$

The rest energy of the particle is mc^2, and (7.4) shows that the ratio of these two energies is close to unity.

The terminology used here (mass, radius, energy of the universe) is not very precise but highly descriptive. Should the universe be finite, then its actual mass, radius and energy would probably be related by a number of order unity to the expressions here used.

Even if the universe is infinite the expressions have some significance as being characteristic of the observable universe or at least that part of it (to use a conventional limit) that shows a Doppler shift of less than, say, $\frac{1}{2}$.

In any case the expressions (7.1)–(7.4) can be constructed in an unambiguous manner from observed well-defined quantities, without regard to their possible interpretations. It is important to remember that the observations used in (7.1)–(7.4) come from such widely separated fields of study as atomic research (e, m_e, m_p), macrophysics (γ, c) and astronomy (ρ_0, T).

It will be seen that the expressions (7.1), (7.2) and the square root of expression (7.3), equal three very large pure numbers differing from each other only by factors of order unity. This is confirmed by relation (7.4), which can be constructed from the others, since (7.4) = (7.3)/(7.1) (7.2).

In considering the possible significance of these near coincidences it should be realized that in their construction all simple numerical factors of order unity (or ten) have been omitted. Such factors as π or 2 or 4π could easily have been used to make the coincidences exact, but speculation on their employment is not likely to be useful in view of the uncertainty of the data.

The likelihood of coincidences between numbers of the order of 10^{40} arising for no reason is so small that it is difficult to resist the conclusion that they represent the expression of a deep relation between the cosmos and microphysics, a relation the nature of which is not understood.

If we admit that some such connexion exists then it is clear that, even more primitive than this quantitative relation, there must also be a qualitative effect of the particle structure of matter on the universe. Most cosmological theories are macroscopic in the sense that they treat matter as homogeneous. This can only be a very rough treatment, but it may be, of course, that any attempt to consider the universe without taking account of the granular structure of matter must be completely unsuccessful at the outset. Although it would be wrong to start from such a pessimistic point of view it seems legitimate to ask whether in the case of any macroscopic theory of cosmology there are any logical (as opposed to mathematical) obstacles to introducing the atomic structure of matter into the theory. In any case it is clear that the atomic structure of matter is a most important and significant characteristic of the physical world which any comprehensive theory of cosmology must ultimately explain.

7.2. There is a further ill-understood branch of microphysics that may be closely related to some aspect of cosmology: the study of cosmic rays. The origin of this very penetrating radiation is at present unknown, but there is a variety of theories, none widely current, some of which connect the rays with random motions within the galaxies, while others consider a stellar origin of the rays more probable (especially in supernovae). Should it, however, prove impossible to ascribe their origin to stars or other galactic objects then it would certainly be an important task of cosmology to account for them. This is especially true if the density of cosmic rays in intergalactic space were as great as it is at the Earth. For then the average energy density of cosmic rays would substantially exceed that of other radiation, though it would still be much less than the energy density of matter.

PART 3

COSMOLOGICAL THEORIES

CHAPTER VIII

THEORETICAL CONCEPTS

8.1. It will be remembered that in Part I of this book two possible approaches to cosmology were discussed. These two approaches lead to two widely differing types of theories. One type is an attempt to extrapolate the physical knowledge gained in terrestrial surroundings and in our astronomical neighbourhood to the universe as a whole, assuming, to start with, the cosmological principle purely as a simplifying hypothesis. The alternative type postulates certain general properties of the universe in the form of a cosmological principle, possibly together with some other axioms as a basis, and then attempts to deduce observable properties of the universe as a whole and of its parts from these postulates. The cosmologies of the general theory of relativity belong to the first type and are discussed in Chapter X, while the very similar Newtonian analogues are described in Chapter IX. Kinematic relativity (Chapter XI) was historically the first representative of the second type of theory, which now also includes the steady-state theory (Chapter XII). This has a relativistic counterpart (12.7) which might be counted as a theory of the first type. The theories of Eddington, Dirac and Jordan, respectively (Chapter XIII), also belong chiefly to that class of theory but draw more on microphysics than does the general theory of relativity.

Owing to the fact that the cosmological principle, in some form or other, enters all these theories, there is a certain body of common concepts and language which will be presented in this chapter.

Although the cosmological principle is only statistical in fact, it becomes most helpful in constructing a theory if it is used as though it applied in detail. The theoretical construction which arises out of this assumption is known as a simple model of the universe or a 'substratum'. Any particle moving with the substratum is called a 'fundamental particle'.

The degree of reality assigned to such a simple model differs according to the theory adopted; it may be regarded as an idealization, a sort of bird's-eye view of the universe in which the detail is 'smeared out', or it may be taken to be the ever-present background or ether against which local irregularities must be considered. In either case the construction of a substratum is the first task of any theory. The origin of the actual irregularities must of course also be explained, but this forms a further task of the theory, in some cases made difficult by mathematical complexities. These explanations must somehow form the link between cosmology and ordinary macroscopic physics, which, it is generally admitted, deals quite satisfactorily with macroscopic problems.

In order to apply the cosmological principle to the construction of the substratum it is necessary to compare the physical aspect of the universe from various points within it. For this purpose imaginary observers have to be introduced, present at all points of space-time. That this procedure is really necessary is easily seen. The main point of the cosmological principle is to assert that our observations are not singular, but typical. What significance can be attached to this assertion other than that any observers like ourselves anywhere would obtain equivalent results from equivalent observations?

The state of motion of these equivalent (or so-called 'fundamental') observers is a matter of considerable importance, revealing a sharp distinction between cosmology and ordinary macrophysics. In particular, in the restricted theory of relativity, equivalent observers are introduced and there they form a *six-parameter family*, any one such observer being specified by his position at a given epoch and by his velocity. The construction of this six-parameter family leads directly to the concept of Lorentz invariance which is of such fundamental importance to physics.

In cosmology, however, there is only a *three-parameter family* of equivalent observers, only one observer passing through each point of space at any given time. *The velocity of an observer is no longer arbitrary, but is a definite function of his position.* It is not difficult to see how this vital property of cosmological theory arises. For imagine an observer passing the Earth at considerable speed. Then to him the universe would present a very different aspect from the one it presents to us. He would see no isotropy, but the nebulae towards which he is moving would look brighter and more violet than the nebulae he is leaving behind. In front he could see some nebulae which are, owing to their red shift, invisible to us, whereas in the rear he could not see some nebulae visible to us. The universe would therefore not present the same aspect to him as it does to us; he would not be an equivalent observer.

A 'fundamental' observer may therefore be specified as one to whom the universe looks isotropic, or, more generally, to whom it presents no first harmonic deviation from isotropy. A more theoretical definition would be that a fundamental observer partakes of the motion of the substratum, that is, he is located on a fundamental particle.

The distinction between special relativity and cosmology is due to the fact that in cosmology the equivalence refers to actual motions, whereas in special relativity it only refers to laws of motion.

The precise way in which these fundamental observers are introduced varies somewhat between different theories, but their distribution and general characteristics are always the same. An argument given by Milne which appears to be of rather restricted applicability shows that the observers must form a continuous rather than a discrete set, but in any case every current theory assumes them to form a continuous set.

The cosmological principle may now be stated as: Every member of the three-parameter set of fundamental observers obtains the same result of corresponding observations of the universe as every other member.

8.2. Up to this point the various theories are in complete agreement about these observers, but the interpretation of the phrase 'observers like us' is a more contentious matter. Our own observations are in many ways so indirect and their interpretation so involved that it is difficult to describe them in simple theoretical terms. A brief description of the various items of equipment of the fundamental observers postulated by the different theories may not be out of place here.

(i) *A clock.* There is universal agreement that a piece of time-keeping mechanism is necessary. General relativity considers that any type of clock would be appropriate, but, according to Milne, different types of clock (atomic, dynamic, nuclear, etc.) might well differ significantly over long periods.

(ii) *A rigid ruler as scale of length.* This seems to be somewhat out of place, since astronomical distances are measured by optical methods and not by rigid rulers, which by definition connect events outside each other's light cones. In the foundation of general relativity a rigid ruler does not seem to be required and is certainly unnecessary for the construction of its cosmological models. Kinematic relativity strongly denies that a rigid ruler is a fundamental measuring device.

(iii) *Light.* Not only do all theories agree that light-receiving apparatus is absolutely essential, but they also agree that, locally at least, the propagation of light is adequately and correctly described by the laws of special relativity. They all agree that our information about the universe is mainly derived from light, but this may be said to be a technical point; it would be imaginable that the bulk of our information about the galaxies might be derived from, say, corpuscular radiations.

(iv) *Theodolites.* The angle between two light rays arriving simultaneously may be measured by such instruments.

(v) *An inertial base.* The existence of such a base allows comparison of angles between light rays arriving at different times. It means that the observer can distinguish between being in a rotating and being in a non-rotating state. This is implicitly accepted in all theories.

(vi) *Measurement of distance.* This is a very difficult point. In practice all measurements of the distances of extragalactic nebulae are carried out by measuring the intensity of the light arriving from them or some stars in them. The rate at which light is emitted by the object in question is then judged by some other criterion (period-luminosity relation for Cepheids or knowledge of average luminosity of the class of object concerned), and the distance is inferred from a comparison of the intensities emitted and received. Distance defined by this method of measurement is known as luminosity distance.

It would seem natural and correct to base a theoretical treatment on this definition of distance. Unfortunately, this has so far proved impossible, and in every theory distance is first defined in a different way. The consequences of such an arbitrary definition are then worked out and the luminosity distances are evaluated at a late stage preparatory to the comparison with observation. This procedure seems to be very unsatisfactory, but at present nothing better appears to be available. Different definitions are used by general relativity and by kinematic relativity, both mathematically simple, but both far removed from practical methods.

General relativity bases itself on the concept of the rigid ruler which enters into its fundamental assumption of the metric. As will be seen later this concept leads in cosmology to the mathematically well-defined but physically somewhat nebulous picture of the 'absolute distance' between 'simultaneous events'.

This measurement of intergalactic distances with rigid rulers is much further removed from physical practice than the definition of distance adopted by Milne as fundamental for kinematic relativity, which is, at least in principle, capable of being carried out. Milne proposed that an observer, in order to measure the distance of a second observer, should send out a light pulse, and that the distant observer should respond by sending out a similar pulse as soon as he receives the first one (or, alternatively, by reflecting it). The first observer could then receive this pulse, and would define the distance of the second observer in terms of the lapse of time, measured by his own clock, between the emission of the first and the reception of the second pulse. This 'radar' method of distance measurement is in practical use for small distances, but its extension to distant astronomical objects is forbidden by the tremendous waiting times involved, by the absence of any responding beings, the swamping of any signals by the radiation of the objects themselves, and by the extreme weakness of the reflected wave. Although these are only objections of technique and not of principle they are very real nevertheless. The first objection in particular is extremely serious, because if these waiting times were at our disposal then our knowledge as obtained by our usual methods would be very great and would vastly exceed what can be obtained by this method.

It was shown by Robertson (1935, 1936) and Walker (1936) that if the cosmological principle is supposed to apply in detail then the metric of relativistic cosmology can be obtained by the light-signal technique without the use of rigid measuring rods.

(vii) *Relativity transformations.* Unless the theory is confined to the special case (which does not correspond to reality) in which the different observers are at rest relatively to each other, allowance must be made, in comparing their observations, for their relative motions. Hence it must be known how the results of observations depend on the observer's state of motion. This is precisely the subject-matter of any theory of relativity and explains the close connexion between such theories and cosmology. Unfortunately, the direct experimental knowledge of the subject is not very extensive. What is known in this way is responsible for and in agreement with the special theory of relativity which is restricted to a particular (inertial) set of relatively moving, but unaccelerated, observers. The well-known Lorentz transformations give the laws of transformation in this case.

All current cosmological theories assume the *local* validity of the velocity addition law of special relativity, but whether they can be used to compare distant observers depends on whether there are accelerations within the substratum. Kinematic relativity denies the existence of such accelerations by a hidden axiom and accordingly takes the Lorentz transformation (derived from its own axioms) to be valid between fundamental observers. In general relativity the question is left open, whereas in the steady-state theory the existence of these accelerations follows from the axioms.

(viii) *Cosmic time.* The Newtonian concept of the uniform omnipresent even-flowing time was shown by special relativity to be devoid of physical meaning, but in 1923 H. Weyl suggested that the observed motions of the nebulae showed a regularity which could be interpreted as implying a certain geometrical property of the substratum (further discussed in Chapter x). This in turn implies that it is possible to introduce an omnipresent *cosmic time* which has the property of measuring *proper time* for every observer moving with the substratum. In other words, whereas special relativity shows that a set of arbitrarily moving observers could not find a common 'time', the substratum observers move in such a specialized way that such a public or cosmic time exists.

Although the existence of such a time concept seems in some ways to be opposed to the generality, which forms the very basis of the general theory of relativity, the development of relativistic cosmology is impossible without such an assumption. Cosmic time is in any case necessary if the ordinary, but not the perfect, cosmological principle is assumed. For if it is asserted that every fundamental observer sees a changing universe, but that it presents the same aspect to them all, then it must be possible for observer A to find a time t_A according to his clock at which he sees the universe in the same state as observer B sees it at a time t_B by his clock. The universe itself therefore acts as a synchronizing instrument which enables A and B (and hence all observers) to synchronize their clocks. Again, by the cosmological principle, if the clock of one of these observers measures his proper time, then all these clocks will measure the proper time of their owners. In this way a universal or cosmic time can be set up.

It is clear from this that kinematic relativity also has a cosmic time, and that cosmic time can only be dispensed with in a theory which adopts the perfect cosmological principle, i.e. in the steady-state theory.

A useful mathematical concept in any model in which a cosmic time exists is the *world map*. This is the distribution of the physical characteristics of the model on the hypersurface on which the cosmic time is constant. The *world picture*, on the other hand, is the aspect of the universe presented to an observer at one instant of time. He sees the distant parts of the universe at an earlier value of the cosmic time than he sees the parts near to him. The highly descriptive names 'world map' and 'world picture' are due to Milne.

It can be shown from fairly simple geometrical arguments, with one or two reasonable assumptions, that a cosmic time must exist in any simple model. This raises the interesting question whether it also exists in the actual universe (as originally surmised by Weyl). It may be that the irregularities and peculiar velocities of the nebulae do not in fact upset the integrability condition which forms the basis of cosmic time. This would be equivalent to a certain law of motion for the actual nebulae, not only for their idealized motions, and in this case a theory based on the existence of a cosmic time in the substratum would apply. However, should this not be the case, i.e. if the peculiar motions of the nebulae are such as not to admit of a cosmic time, then it seems very likely that a theory of the substratum based on the existence of a cosmic time does not even approximately apply to the actual universe. For then the construction of a substratum is an idealization of the actual universe which admits of something which has no counterpart in the general case, namely a cosmic time. This point may be of some interest for evolutionary theories since these necessarily involve a cosmic time. The question is common to all theories other than the steady-state theory, but very little attention seems to have been paid to resolving this problem which is relevant to the logical foundations of all these theories.

(ix) *Duration of observations.* As a further point the duration of the observations of these idealized fundamental observers may be discussed. Although the length of time during which actual cosmological observations have been carried out is extremely short compared with any estimate of the time-scale of the universe, there is, nevertheless, a time measurement involved. For any observation of the Doppler shift takes a short but finite time, since it involves a comparison of the periods of two oscillations. Our observations cannot therefore be called strictly instantaneous; but the minimum length of time they necessarily occupy (about 10^{-14} sec.) is so very short compared with a time in which the distance of a nebula could be expected to alter appreciably (about 10^{8} years) that it seems permissible to classify ideally possible observations as 'rapid' and 'slow' respectively. Measurements of apparent luminosity, Doppler shift, number counts, etc., are 'rapid', whereas measurements of nebular distances by observing the lapse of time between emission of a ray and the reception of its reflexion are 'slow'. The distinction, it must be emphasized, is purely one of degree, but may nevertheless be important. Certain theoretically predicted effects may be 'slow' and therefore can only be noticed by sufficiently 'long-lived' observers, whereas others of more immediate interest and, in a sense, of more reality, are accessible to 'short-lived' observers like ourselves.

In this connexion a few remarks may be made concerning the question that is sometimes asked, whether the recession of the nebulae is 'real'. The question has been discussed by Hubble and Tolman (1935) and by Hubble (1936), and they have taken the view that if the recession is not 'real' its effect on the intensity of light received should be less important than if the recession is 'real'. In general relativity (the theory on which these authors based their work) an 'unreal' recession is impossible, but in kinematic relativity some meaning may be ascribed to this term (p. 138).

The question is evidently meaningless without a definition of distance. Since the only natural definition is that of luminosity distance, it seems that the form of greatest physical significance that can be given to this question is whether there is any progressive dimming of the light of a distant nebula in addition to any effects due to a variation of Doppler shift or intrinsic luminosity. In this form, then, the question is whether there are any 'slow' effects of the 'rapidly' measurable Doppler shifts.

(x) *Creation.* One of the most fundamental laws of terrestrial physics is that of the conservation of mass (energy). The applicability of this law to cosmology forms a very debatable question on which the various theories disagree.

The law itself is clearly an extrapolation from experience. Laboratory experiments show that matter is conserved, at least to experimental accuracy, and for all terrestrial applications this accuracy is amply sufficient. On the other hand, terrestrial regions and periods of time are not typical from a cosmic point of view. The densities are far greater than average cosmic densities, and the periods of time are very short on the cosmic scale. Accordingly, a rate of creation of matter per unit volume per unit time, which is far too low to be detected experimentally, could nevertheless be of the greatest cosmic significance. It follows that terrestrial experience is no guide whatever in assessing whether there are processes of creation or annihilation going on at a rate significant in cosmology, and the question must be left open.

Terrestrial physical theories are based on the law of conservation of mass. If it is considered that recourse should be had to such theories in constructing a cosmology, then, in the absence of any theory not postulating conservation of mass, the law has to be carried over into cosmology. This step is taken in the cosmologies of Newtonian physics and in general relativity. In kinematic relativity, although it is not based on local physical theories, the law of conservation of mass (particle number) is introduced as an axiom.

The scope of any such theory postulating conservation of mass is necessarily somewhat restricted. For by this very postulate the theory declares itself unable to discuss the problem of the origin of matter in the universe. This point was put most strongly by Milne, who considered that this problem does not appertain to science. Briefly, in his view, any question as to the origin of matter is a question connected with the zero of time, and, since in his view science can only deal with epochs $t > 0$, any such question is unscientific. Although the point of view of general relativity has not been put so clearly and precisely, it cannot in fact be very different. In the view of all these theories the question of the origin of matter is as meaningless as, say, the question of the simultaneous precise determination of the position and momentum of an electron in quantum theory.

Any theory which permits creation is immediately in a very different position. The scope of its inquiry is not limited by what must appear to be an artificial restriction. The creation of matter is then no longer a subject which has to be accepted by science without query in blind faith, but becomes a new field in which scientific methods can be successfully applied. The rules governing the creation process, although at present not detectable by direct means, will lead to consequences of cosmological interest. Some of these results can be proved or disproved by observation, and so the details of the creation process should gradually become known by indirect methods.

It is very much in the spirit of scientific inquiry to welcome any theory which widens the range of applicability of science. In the case discussed the gain is very great, but it has to be purchased at the price of rejecting the applicability to cosmology of locally established theories. To a greater or lesser extent any creation-type theory of cosmology must be based in the first instance on cosmological considerations rather than on established physical theories.

CHAPTER IX

NEWTONIAN COSMOLOGY

9.1. The first attempts to grapple with the cosmological problem as a whole were made during the nineteenth century, when the framework of Newtonian theory was accepted without question. It is a curious fact that these attempts were destined to fail not because of their Newtonian origin but because, in addition to the cosmological principle, the further assumption was made that the universe was static in the sense that there were no large-scale motions of matter. It will be seen in the course of this section that under these conditions there can be no solution in strictly Newtonian terms. This failure led to a diminution of interest in cosmology that lasted until Einstein began the exploration of the cosmological consequences of general relativity in his famous paper of 1916. The next fifteen years were occupied with the further development of relativistic cosmology, including non-static models, to the total exclusion of Newtonian cosmology. It was only in 1934 that Milne and McCrea reverted to the Newtonian problem and showed that, with a suitable interpretation of the Newtonian terms, Newtonian cosmology was in many respects completely equivalent to relativistic cosmology. This new formulation of the old subject is highly interesting, since in spite of our present denial of many of the premisses of Newtonian theory it reveals many of the essential features of relativistic cosmology without the mathematical complexity. It also makes the significance of the different terms apparent much more readily, and finally it shows a suitable Newtonian formulation to lead to equations close to the relativistic ones not only (as is well known) in ordinary macroscopic physics, but also for large-scale phenomena, provided only that the densities are not vastly in excess of what is known to exist.

In this chapter, then, Newtonian theory is fully accepted, and an attempt is made to construct the largest possible dynamical system. The cosmological principle is assumed essentially as a simplicity postulate. The main subject of the discussion is the motion of the substratum, idealized as the streaming of a uniform fluid. To any observer moving with a particle of the fluid the model presents the same appearance as to any other similarly moving observer provided they carry out their observations at the same time. This is the relevant formulation of the cosmological principle. Owing to the assumptions made there is a uniform even-flowing Newtonian time t and the problem of clock synchronization does not arise (since its solution is implicit in the Newtonian axioms).

9.2. Consider then an observer O moving with the substratum. He sets up a system of coordinates with himself as origin and observes the physical properties at a general point P of the liquid at time t, characterizing the point P by the position vector $\mathbf{r} = OP$. His observations will reveal the velocity $\mathbf{v}$ relative to him of P, which, being a function of $\mathbf{r}$ and t, may be written $\mathbf{v}(\mathbf{r}, t)$. Similarly, he discovers the density $\rho(\mathbf{r}, t)$ at P and the pressure $p(\mathbf{r}, t)$.

Consider now a second observer O' moving with the substratum. By a similar procedure he can find $\mathbf{v}'(\mathbf{r}', t)$, the relative velocity $\mathbf{v}'$ of the point P, whose position vector relative to him is $\mathbf{r}'$ at time t. Correspondingly he finds $\rho(\mathbf{r}', t)$ and $p(\mathbf{r}', t)$, where we have omitted the dashes on ρ and p, since density and pressure are quantities defined independently of the observer. The cosmological principle now demands that $\mathbf{v}'$, ρ and p should be the *same* functions of $\mathbf{r}'$ and t as $\mathbf{v}$, ρ and p are of $\mathbf{r}$ and t. For otherwise O' and O would have different pictures of the universe.

In the next few arguments spatial variations only are considered, so we may, for the moment, omit the variable t. Let the vector OO' be $\mathbf{a}$. Then the velocity of O' relative to O is $\mathbf{v}(\mathbf{a})$, since O' is moving with the liquid. If O and O' are looking at the same particle of the liquid then $\mathbf{r}' = \mathbf{r} - \mathbf{a}$. It follows then that

$$\mathbf{v}'(\mathbf{r}-\mathbf{a}) = \mathbf{v}(\mathbf{r}) - \mathbf{v}(\mathbf{a}), \quad \rho(\mathbf{r}-\mathbf{a}) = \rho(\mathbf{r}), \quad p(\mathbf{r}-\mathbf{a}) = p(\mathbf{r}). \tag{9.1}$$

Now by the cosmological principle $\mathbf{v}'$ is the same function of its argument as $\mathbf{v}$. Furthermore, $\mathbf{r}$ and $\mathbf{a}$ are arbitrary vectors. The second and third of equation (9.1) hence imply that ρ and p should be independent of position, whereas $\mathbf{v}$ must satisfy the functional equation

$$\mathbf{v}(\mathbf{r}-\mathbf{a}) = \mathbf{v}(\mathbf{r}) - \mathbf{v}(\mathbf{a}). \tag{9.2}$$

This shows that $\mathbf{v}$ is a linear vector function of its argument, so that we may write

$$\mathbf{v}(\mathbf{r}) = V\mathbf{r}, \tag{9.3}$$

where V is a tensor independent of $\mathbf{r}$. This tensor may, in the usual way, be divided into a symmetrical and an antisymmetrical part. The latter corresponds to a rigid-body rotation of the universe around the observer. The angular velocity of this rotation could be altered if the observer chose a new system of reference rotating relatively to the old system. Accordingly, this rigid-body rotation represents a property of the observer's frame of reference, and not of the universe. For simplicity (and in accordance with observational procedure) we may assume that our observers always choose the motions of their frames of reference so as to reduce this rigid-body rotation to nil. This is a pure convention and not a restrictive assumption. It ensures that V is a symmetrical tensor. Furthermore, the observer will be able to set up his system of coordinates coincident with the principal axes of V, so that

$$v_x = Ax, \quad v_y = By, \quad v_z = Cz, \tag{9.4}$$

where A, B and C are the diagonal elements of the tensor. In other words, there will be a velocity-distance relation in operation in that each component of the velocity of a body is proportional to its corresponding coordinate, but the three factors of proportionality are not necessarily equal. This is as far as one can go on the assumption of homogeneity; but if, as is usual, isotropy is also assumed then $A = B = C$. In this simple case the results become (by restoring the explicit time dependence of the quantities)

$$\mathbf{v} = f(t)\,\mathbf{r}, \tag{9.5}$$

$$\rho = \rho(t), \tag{9.6}$$

$$p = p(t). \tag{9.7}$$

If (9.5) is regarded as the equation of motion of a particle it may be integrated in the form

$$\mathbf{r} = R(t)\,\mathbf{r}_0, \tag{9.8}$$

where $R(t)$ satisfies

$$\frac{1}{R}\frac{dR}{dt} = f(t), \quad R(t_0) = 1, \tag{9.9}$$

and $\mathbf{r}_0$ is the position vector of the particle at time t_0. This shows that the only motions compatible with homogeneity and isotropy are those of uniform expansion or contraction, a simple scaling up or down with a time-dependent scale factor.

This completes our examination of the kinematics of the motion of the substratum. The next step must be the imposition of the principles of conservation of mass and momentum in the form of the equation of continuity and Euler's equations of motion.

By the equation of continuity and by (9.5) and (9.6)

$$0 = \frac{\partial \rho}{\partial t} + \operatorname{div}(\rho \mathbf{v}) = \frac{d\rho}{dt} + 3\rho(t) f(t). \tag{9.10}$$

This may be integrated by virtue of (9.9)

$$\rho(t) = \frac{\rho(t_0)}{R^3(t)}. \tag{9.11}$$

This is really an obvious consequence of (9.8). If all linear dimensions are scaled up by the factor $R(t)$, then all volumes are increased by $[R(t)]^3$ and the densities must be correspondingly diminished.

9.3. Before imposing Euler's equations we have to investigate whether O's system of coordinates is inertial. This involves the calculation of the gravitational field which is not unambiguous in an unbounded system. A definitive method is to apply the Newtonian considerations to a bounded system (Layzer and McCrea, 1954) such as an arbitrarily large spherical body of matter in otherwise empty space, the local application of the cosmological principle ensuring homogeneity throughout the interior of the body. It can be shown that, if the central particle is inertial, in the strict sense of the word, and if by the assumption of isotropy rotation is excluded, then the combination of inertia and gravitation implies that, as far as local dynamics goes, each interior observer may regard himself as inertial.

It is interesting to note that every observer's system is inertial although different observers may be accelerated relatively to each other. This is of course not permissible according to the strictly Newtonian system, but as long as we assume that each observer only uses his own system no difficulties or contradictions arise.

Euler's equations become, by (9.5), (9.6) and (9.7),

$$\frac{D\mathbf{v}}{Dt}+\frac{1}{\rho}\operatorname{grad}p-\mathbf{F}=\mathbf{r}\left[\frac{df}{dt}+f^2\right]-\mathbf{F}=0, \tag{9.12}$$

where **F** is the body force per unit mass, i.e. gravitation. The full evaluation of **F** in an infinite system is a somewhat ambiguous matter,* but for our purposes it is sufficient to use Poisson's equation

$$\operatorname{div}\mathbf{F}=-4\pi\gamma\rho. \tag{9.13}$$

Taking the divergence of (9.12) we have therefore

$$3\left(\frac{df}{dt}+f^2\right)=-4\pi\gamma\rho. \tag{9.14}$$

This result could have been deduced directly from (9.12) if it had been assumed that the effective gravitational force on a particle P viewed from O is due entirely to the sphere of matter with its centre at O and its surface passing through P. Then

$$\mathbf{F}=-\frac{4\pi}{3}\gamma\rho\mathbf{r}. \tag{9.14a}$$

Substituting (9.9) and (9.11) in (9.14) we have

$$R^2\frac{d^2R}{dt^2}+\frac{4\pi}{3}\gamma\rho(t_0)=0. \tag{9.15a}$$

It will be seen from (9.15a) that a static universe ($R=1$) is impossible except in the trivial case when the density vanishes. This led to the difficulties of cosmology in the nineteenth century briefly referred to above. It was then proposed to overcome these difficulties by a change in the Newtonian law of gravitation which would become appreciable only at very large distances without serious effects within the solar system. The changes then proposed did not find much favour owing to the arbitrariness involved. In the general theory of relativity an equation exactly analogous to (9.15a) occurs, and in that theory the corresponding alteration of the law of gravitation is not arbitrary, but for certain mathematical reasons it can only be of one form unless very great complications are introduced. The Newtonian analogue of this relativistic procedure is also very well defined and consists of introducing in (9.13) a term

* Recently elucidated by D. Layzer and W. H. McCrea (1954).

proportional to the distance and independent of the density. This term serves the same purpose as that introduced on a Newtonian basis by Neumann (1896) and Seeliger (1895, 1896).

Since the chief purpose of our discussion is to develop the Newtonian counterpart of relativistic cosmology, we shall introduce this hypothetical term into our discussions by altering (9.14*a*) to read

$$\mathbf{F} = -\frac{4\pi}{3}\gamma\rho(t)\,\mathbf{r}+\tfrac{1}{3}\lambda\mathbf{r}, \tag{9.14b}$$

where λ, the so-called cosmological constant, has the dimensions $(\text{time})^{-2}$, and the factor $\frac{1}{3}$ has been introduced for later convenience. Equation (9.15 *a*) then becomes

$$R^2\frac{d^2R}{dt^2}+\frac{4\pi}{3}\gamma\rho(t_0)-\tfrac{1}{3}\lambda R^3 = 0. \tag{9.15b}$$

Although the cosmological term was at first introduced in order to obtain a static model (the requisite value for λ is evidently $4\pi\gamma\rho(t_0)$) all the solutions of (9.15 *b*) for various values of λ have attracted great interest since. Except for the physical interpretation of various quantities the integrated form of (9.15 *b*)

$$\left(\frac{dR}{dt}\right)^2 = \frac{C}{R}-k+\tfrac{1}{3}\lambda R^2 = G(R) \quad (\text{say}), \tag{9.16}$$

is identical with the corresponding equation of general relativity. The quantity k is a constant of integration which is of dimension $(\text{time})^{-2}$ like λ and $\gamma\rho(t_0)$, and, for brevity, we have put

$$8\pi\gamma\rho(t_0) = 3C.$$

9.4. Equation (9.16) may be integrated in terms of elliptic functions, but it will be sufficient for our purposes to carry out a qualitative integration.

A number of different cases must be distinguished. Some of these may be arrived at from others by reversing the direction of time which is always arbitrary. In order to eliminate this trivial duplication the direction of time will be chosen, where possible, so as to lead to expansion (R increasing). The origin of time is also arbitrary and will be chosen later on for convenience.

The parameters λ and k in (9.16) are quite arbitrary, but since we are only concerned with positive mass densities, C must be positive. The following main cases and subdivisions then arise (see also Figs. 1–4):

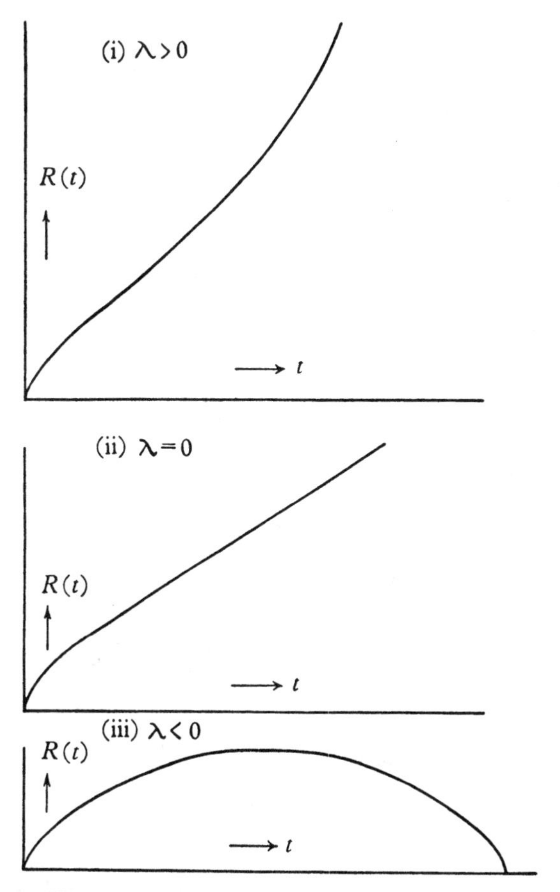

Fig. 1. World models of Newtonian and relativistic cosmologies. Case 1: $k<0$.

Case 1: $k<0$.

(i) $\lambda>0$. In this case $G(R)$ is a positive function of R, with a minimum at $R_m=(3C/2\lambda)^{\frac{1}{3}}$. For small t, $R\sim(9Ct^2/4)^{\frac{1}{3}}$. The rate of expansion slows down until $R=R_m$ and then increases again. For large t, $R\sim\exp[t(\frac{1}{3}\lambda)^{\frac{1}{2}}]$.

(ii) $\lambda = 0$. $G(R)$ is now a positive decreasing function of R. For small t we again have $R \sim (9Ct^2/4)^{\frac{1}{3}}$. The rate of expansion slows down continuously. For large t, $R = t(-k)^{\frac{1}{2}} + O(1)$.

(iii) $\lambda < 0$. $G(R)$ is a decreasing function of R, positive in $0 \leqslant R < R_c$ and negative for $R > R_c$, where R_c is the root of a certain cubic equation. The expansion begins as in the preceding cases, but slows down until R reaches its maximum value R_c. Then contraction sets in, the system runs through its previous phases in the opposite sense until $R = 0$, when the cycle starts again. This is an oscillating model.

Case 2: $k = 0$.

(i) $\lambda > 0$. $G(R)$ is positive, with a minimum at $R_m = (3C/2\lambda)^{\frac{1}{3}}$, just as in Case 1 (i). The entire behaviour is very similar to that case except that here there is the explicit solution

$$R^3 = \frac{3C}{2\lambda}[\cosh\{t(3\lambda)^{\frac{1}{2}}\} - 1].$$

(ii) $\lambda = 0$. This is the simplest of all cases, the solution being

$$R = (\tfrac{9}{4}Ct^2)^{\frac{1}{3}}.$$

(iii) $\lambda < 0$. The case is very similar to Case 1 (iii), but here there is the explicit solution

$$R^3 = \frac{3C}{2(-\lambda)}[1 - \cos\{t(-3\lambda)^{\frac{1}{2}}\}].$$

Case 3. $k > 0$.

This case shows the greatest variety. There is a critical value λ_c of λ defined by

$$\lambda_c = \frac{4k^3}{9C^2}.$$

(i) $\lambda > \lambda_c$. $G(R)$ is positive, with a minimum at $R = [3C/2\lambda]^{\frac{1}{3}}$. This case is similar to Cases 1 (i) and 2 (i).

(ii) $\lambda = \lambda_c$. $G(R)$ is positive except that $G(R) = 0$ has a double root at $R = 3C/2k = R_c$ (say).

(ii*a*) Accordingly, there is a static solution $R = R_c$. There are two other solutions:

(ii*b*) $R \sim (9Ct^2/4)^{\frac{1}{3}}$ at first, then expansion slows down and R approaches R_c asymptotically from below.

(ii*c*) R approaches R_c asymptotically from above as $t \to -\infty$. Expansion accelerates all the time, and, for large t,

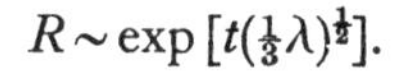

$$R \sim \exp [t(\tfrac{1}{3}\lambda)^{\frac{1}{2}}].$$

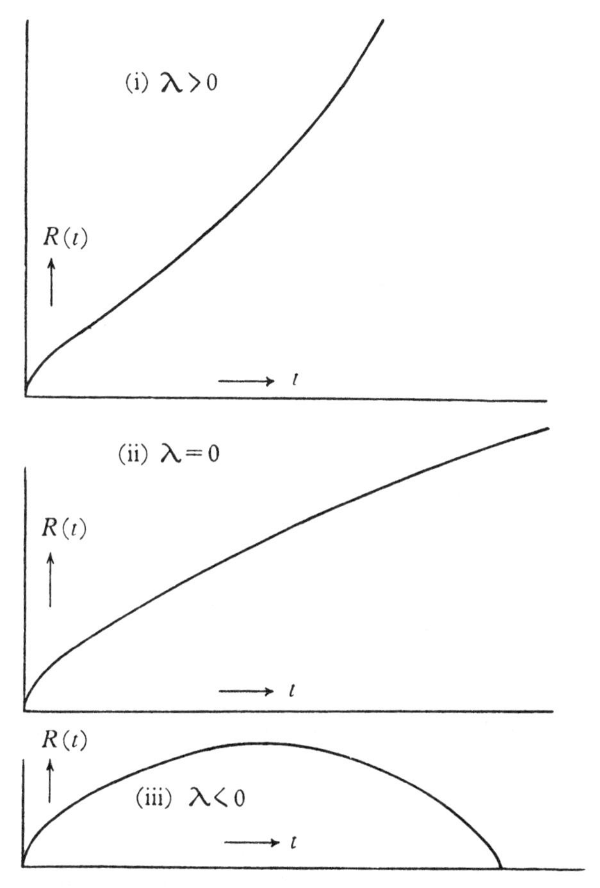

Fig. 2. World models of Newtonian and relativistic cosmologies. Case 2: $k = 0$.

(iii) $\lambda_c > \lambda > 0$. $G(R)$ is positive for sufficiently large and for sufficiently small values of R, but there is a stretch (say $R_1 < R < R_2$) where $G(R)$ is negative. Hence there are two solutions:

(iii*a*) In this case $0 \leqslant R \leqslant R_1$. The solution is very similar to Cases 1 (iii) and 2 (iii), i.e. we have an oscillating universe.

(iii$b$) $R_2 \leqslant R$. R varies like $\exp[(-t)(\frac{1}{3}\lambda)^{\frac{1}{2}}]$ as t tends to $-\infty$. The rate of decrease slows down until R reaches its minimum value R_2. After that, expansion sets in and accelerates until for large t

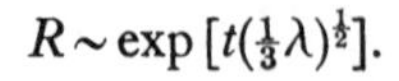

$$R \sim \exp[t(\tfrac{1}{3}\lambda)^{\frac{1}{2}}].$$

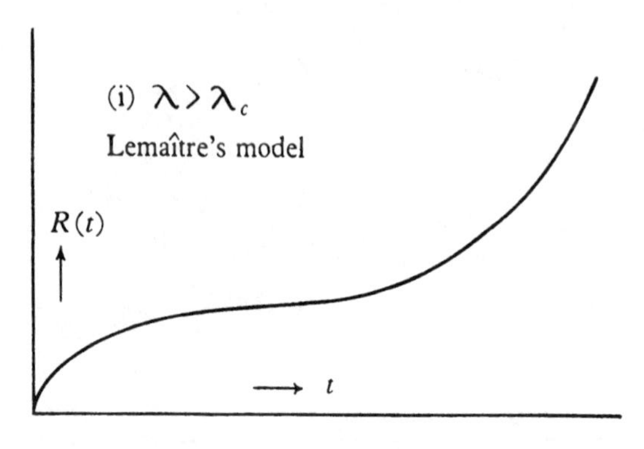

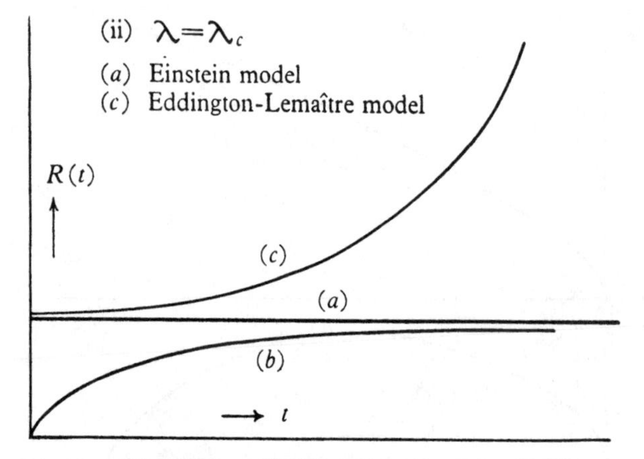

Fig. 3. World models of Newtonian and relativistic cosmologies. Case 3 : $k > 0$. Subclasses (i) and (ii).

(iv) $0 \geqslant \lambda$. $G(R)$ decreases monotonically. The case is similar to Cases 1 (iii) and 2 (iii).

This enumeration covers all possible cases. They may be classified by the type of model of the universe they lead to in the following way:

Class I. The static or Einstein universe: Case 3 (ii$a$).

Class II. Monotonically expanding models starting at a definite time from a point origin $R = 0$: Cases 1 (i), 1 (ii), 2 (i), 2 (ii), 3 (i). It should be noted that in the (i) cases dR/dt passes through a minimum, but this is not so in the (ii) cases.

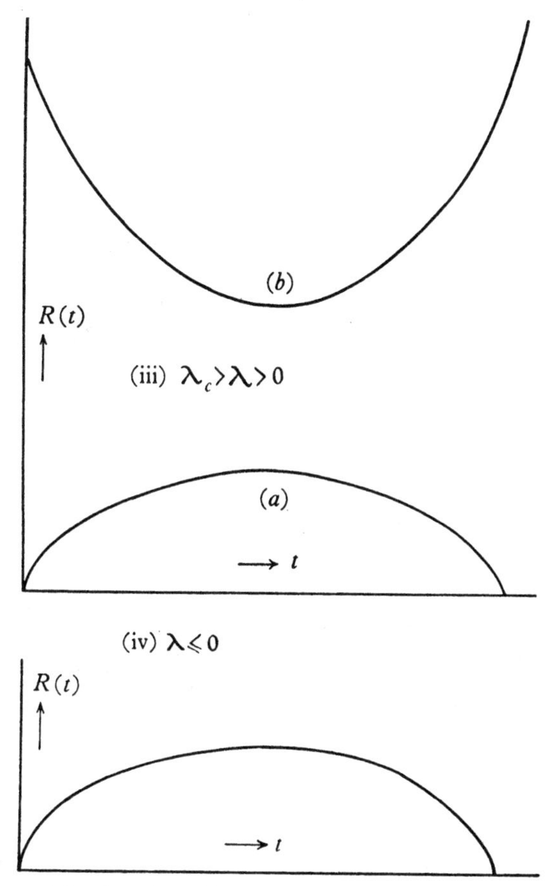

Fig. 4. World models of Newtonian and relativistic cosmologies. Case 3: $k > 0$. Subclasses (iii) and (iv).

Class III. The model which expands from a finite value of R at $t = -\infty$, the rate of expansion increasing gradually. This is the Eddington-Lemaître model: Case 3 (ii$c$).

Class IV. The model starts at a finite time from the point origin $R = 0$ and expands with decreasing speed. R tends to a limit as $t \to \infty$: Case 3 (ii$b$).

Class V. Models which oscillate between $R = 0$ and a finite value of R: Cases 1 (iii), 2 (iii), 3 (iii *a*), 3 (iv).

Class VI. The model which contracts from infinity to a finite value of R and then expands again to infinity: Case 3 (iii *b*).

9.5. We have now obtained complete descriptions of the models, since the run of the function $R(t)$ completely determines the behaviour of the model. However, we have taken rather a bird's-eye view of the problem looking, as it were, at the model from outside. We have to consider now how any of these models would appear to an observer in our position, that is, in the model and partaking of its motion. Since all our observations consist of examinations of the light that arrives from distant objects, the propagation of light in the models has to be considered.

Here the intrinsic weakness of the Newtonian method reveals itself. Owing to the artificial division between dynamics and electromagnetism the dynamical behaviour of the system does not determine the properties of light. Reasonable results can only be obtained by using post-Newtonian concepts and methods.

It follows from the cosmological principle that the *local* velocity of light should be the same for all observers at the same time, although it may be a function of the time. Assuming the Newtonian law for the addition of velocities to hold even if one of these is the local velocity of light $c(t)$, it follows that for incoming and outgoing radial rays, respectively,

$$\frac{dr}{dt} = \frac{\dot{R}}{R} r \pm c(t),$$

which is equivalent to

$$R\frac{d}{dt}\left(\frac{r}{R}\right) = \pm c(t), \qquad (9.17)$$

where the form of the function $c(t)$ cannot be determined by *a priori* reasoning.

Equation (9.17) may be interpreted as expressing the law of propagation of light rays in an ether that is partaking of the expansion of the universe. Clearly only such an ether can be consistent with the cosmological principle, otherwise the local ether wind could be used to distinguish between observers 'really' at rest and observers 'really' in motion. Such a distinction, incompatible with cosmology, would be required for any more direct interpretation of the cosmic Doppler shift, since the Newtonian Doppler-shift formula depends not only on the *relative* velocity of observer and source but on their absolute velocities.

In order to make further progress it is perhaps desirable to assume a special form for the function $c(t)$. Simplicity suggests that it should be taken to be constant, and this assumption has the further great merit of again leading to formulae closely parallel to the relativistic results. Integrating (9.17) we have

$$\frac{r_2}{R_2} - \frac{r_1}{R_1} = \pm c \int_{t_1}^{t_2} \frac{dt}{R(t)}, \tag{9.18}$$

where r_1 has been written for $r(t_1)$, etc.

Consider now a light ray from a source at $r(t)$ travelling to an observer at $r = 0$. Then

$$\frac{r_1}{R_1} = c \int_{t_1}^{t_2} \frac{dt}{R(t)}. \tag{9.19}$$

This equation may be regarded as determining t_2, the time of reception, in terms of t_1, the time of emission, and r_1, the distance of the source at the time of emission. Consider now the *same* source, moving with the substratum, emitting light at various times t_1. Then owing to the motion of the substratum the left-hand side of (9.19) is independent of t_1. Accordingly, if we concentrate attention on two light rays emitted at successive instants t_1 and $t_1 + \Delta t_1$ respectively, and received at t_2 and $t_2 + \Delta t_2$ respectively, then by (9.19)

$$\frac{\Delta t_1}{R_1} = \frac{\Delta t_2}{R_2}, \tag{9.20}$$

and hence

$$\frac{\Delta t_2}{\Delta t_1} = \frac{R_2}{R_1} = \frac{R(t_2)}{R(t_1)}. \tag{9.21}$$

But if Δt_1 represents the period of the emitted wave then Δt_2 represents the period of the wave received, and so in consequence of the constancy of the velocity of light here assumed (9.21) represents the Doppler shift $1 + \delta\lambda/\lambda$. This formula, too, is identical with the relativistic formula.

Confining our attention, for definiteness, to the case in which $R(t)$ is an increasing function, then (9.21) expresses not only the fact that the light received appears to be reddened but also that it is weakened in intensity. A strict derivation of the intensity reduction formula can only be given on the basis of relativity, but the following argument, though by no means rigorous, gives the analogous answer. Imagine the source to be sending out light quanta (photons). Then these quanta will appear to the observer to be weakened in energy by the Doppler-shift factor R_1/R_2, since the frequency of the light has been reduced and Planck's constant is, in this picture, assumed to be constant in time. Furthermore, the arrival rate of the quanta is similarly reduced by the same factor, since the time between the reception of successive quanta is R_2/R_1 times as large as the interval separating their emission. Accordingly, in addition to the geometrical weakening, the intensity is reduced in the ratio $(R_1/R_2)^2$. (In a contracting universe there would clearly be a violet shift and an increase in intensity.)

Consider now an isotropic source whose distance from the observer, supposed at the origin, was r_0 at the time when $R = 1$, and suppose the rate of emission of light by this source is $L(t_1)$. Then at t_1, the instant of reception, the light has spread isotropically around the source over a sphere with the source at the centre and its surface passing through the observer. Accordingly, the radius of this sphere is $r_0 R(t_2)$. Hence, allowing for the factor considered above, the intensity received is

$$l = \frac{L(t_1)}{4\pi r_0^2 R_2^2} \frac{R_1^2}{R_2^2} = \frac{L(t_1) R_1^2}{4\pi r_0^2 R_2^4}. \tag{9.22}$$

Usually $L(t)$ is taken to be a constant.

Similarly, if the assumption is made that no new nebulae form, at least within the periods under consideration, then the number N of nebulae whose distance was less than r_0, when R was equal to unity, satisfies

$$N = br_0^3, \tag{9.23}$$

where b is a constant.

Finally, by (9.19), r_0 and t_1 are connected, for a given instant of observation t_2, by

$$r_0 = c\int_{t_1}^{t_2} \frac{dt}{R(t)}. \tag{9.24}$$

If the function $R(t)$ were known, then (9.21), (9.22) and (9.23), with the aid of (9.24), would establish two relations between the three observable quantities connected with nebulae, namely, their apparent brightness l, Doppler shift $1+\delta\lambda/\lambda$ and number N. Conversely, it should be possible to deduce the run of $R(t)$ if, say, N were known observationally as a function of l. The variation of Doppler shift with l could then be deduced and checked against the observed variation and so give an indication of the validity of the theory.

In fact, of course, Newtonian theory serves only as a picture, though a very useful picture, owing to the close analogy with relativistic cosmology. Many of the formulae of general relativity established by the use of the cumbrous tensor calculus become much clearer when the Newtonian interpretation of the formulae is remembered. Nevertheless, it is hardly worth while nowadays to compare Newtonian theory and observation, since the Newtonian concepts are known to be untenable.

9.6. The main feature of Newtonian theory, which will be met with again in relativistic cosmology, is that by taking a bird's-eye view of the whole universe it can be shown that its whole behaviour is characterized by a single function $R(t)$, which satisfies a certain differential equation (9.16). The existence of a cosmic universal time plays a fundamental part in the theory. The observable relations are only deduced at the very end when the model has been constructed, and are of mathematically somewhat awkward form.

One of the most important consequences of this discussion is that it shows that there is no limit to the size of object to which Newtonian ideas can be applied. The divergencies from the relativistic results are minute provided the *local* velocities are small compared with the velocity of light, and the *local* gravitational potential energy of matter is small compared with its rest mass. In these circumstances Newtonian theory gives excellent results on the terrestrial scale and in the solar system. The work of Milne and McCrea has shown that this is also true on the cosmic scale. Hence in any problem of intermediate size, such as the structure of nebulae, general relativity cannot be expected to explain any major features in any different or better way than Newtonian theory.

CHAPTER X

RELATIVISTIC COSMOLOGY

10.1. The limitations of Newtonian theory became apparent in the last chapter as soon as the propagation of light was discussed. It was a different case of the same problem, the union of electrodynamics and mechanics which, in the shape of the Michelson-Morley experiment, destroyed the faith in Newtonian theory and led to the formulation of the special theory of relativity. This theory, which has been verified in many ways and whose correctness is not in serious doubt, suffers nevertheless from the severe limitation that it does not possess a fully satisfactory theory of gravitation. Hence it is unable to comprehend the case of mutually accelerated observers which, as will be seen later, is vitally important in cosmology. For these reasons a direct application of the theory to cosmology is not possible, though the theory forms the foundation on which kinematic relativity has since been built.

The two limitations mentioned show clearly that special relativity cannot claim to be as complete as Newtonian theory. An extension of the theory is required, and is provided by the general theory of relativity. This theory covers the entire field of ordinary macroscopic physics, but is in doubt to a far greater extent than special relativity. It is considered to be correct by a majority of theoretical physicists, but there is a substantial minority that considers it to be wrong or, at least, not established. In view of the cosmological consequences it will perhaps not be out of place to give here a brief description of the theory, of its achievements and of the arguments brought against it.

10.2. The fundamental idea of the theory is that it is possible anywhere, at any time, to choose inertial frames of reference in which the laws of special relativity are satisfied in the immediate neighbourhood of the observer though possibly not for regions at some distance from him. For example, it is possible to choose at any point of the Earth's surface a freely falling frame of reference (represented, say, by a freely falling lift). Referred to this frame (i.e. as observed by a man in the lift) gravitation does not exist, since the acceleration due to gravitation is the same for all objects and is cancelled by the acceleration of the lift. This cancellation occurs only if the gravitational and inertial masses of any body are strictly identical, i.e. if Galileo's principle that all bodies fall at the same speed is accurately true. The general theory of relativity incorporates this fact, unexplained in Newtonian physics. The equality of the masses makes it reasonable to assume that special relativity holds in the lift, but if events at a distant point of the Earth's surface were referred to this frame of reference, then owing to the different direction of gravity the gravitational acceleration there would not be cancelled by the lift's acceleration. Hence the validity of special relativity is purely local.

The laws of *special* relativity may be put into the form of the following statements:

(i) An inertial observer locates events by setting up one time coordinate t and three spatial coordinates (say rectangular Cartesians x, y, z). Any other inertial observer can set up similar coordinates $(t'; x', y', z')$, but the same event will correspond to a different set of coordinate numbers in his system from that in the system of the first observer.

(ii) The coordinates assigned to events by different observers are, however, related by the rule that if $(t_1; x_1, y_1, z_1)$, $(t_2; x_2, y_2, z_2)$ are the coordinates of two events in the system of one observer and c is the velocity of light, then the quantity

$$s^2 = (t_1 - t_2)^2 - [(x_1 - x_2)^2 + (y_1 - y_2)^2 + (z_1 - z_2)^2]/c^2 \qquad (10.1)$$

is independent of the observer. Although a different observer will obtain different coordinates he can construct the same quantity and will obtain the same value for it.

If the events are close together then (10.1) may be written in differential form

$$ds^2 = dt^2 - \frac{dx^2 + dy^2 + dz^2}{c^2}. \qquad (10.2)$$

ds is often called the interval.

(iii) The orbits of material particles in the absence of electromagnetic fields (and, since we are now dealing with special relativity, of gravitational fields) are straight lines in the four-dimensional space $(t;\, x,\, y,\, z)$. The slope of these straight lines is subject to the condition

$$c^2 > \left(\frac{dx}{dt}\right)^2 + \left(\frac{dy}{dt}\right)^2 + \left(\frac{dz}{dt}\right)^2.$$

(iv) The tracks of light rays are also straight lines, but satisfy

$$c^2 = \left(\frac{dx}{dt}\right)^2 + \left(\frac{dy}{dt}\right)^2 + \left(\frac{dz}{dt}\right)^2.$$

The laws of the transformation of coordinates (Lorentz transformation) between two inertial observers follow from (10.1) or (10.2). Maxwell's equations of electrodynamics can be written in a form appropriate to equation (10.1) and a system of wave optics (as opposed to the ray optics (iv)) may be obtained, but this will not be required in our discussion.

It should be observed that condition (ii) requires different observers to use the same units of length and time and hence requires the existence of rigid rulers and standard clocks.

One may easily extend this scheme by saying that in the absence of gravitational fields it would be possible for an arbitrarily moving observer to choose four general coordinates (x^0, x^1, x^2, x^3) functionally related to $(t;\, x, y, z)$. He would then find that in spite of his peculiar coordinates he would agree with all other observers about the value of the interval ds separating two neighbouring events. If the functional relations defining x^0, x^1, x^2, x^3 are substituted into (10.2) it is changed into

$$ds^2 = \sum_{\mu=0}^{3} \sum_{\nu=0}^{3} g_{\mu\nu}\, dx^\mu\, dx^\nu, \tag{10.3}$$

where the $g_{\mu\nu}$ are ten functions (since there is no point in distinguishing between $g_{\mu\nu}$ and $g_{\nu\mu}$) of the four coordinates. Now general relativity maintains that (10.3) is always valid in the sense that all observers agree on the value of ds, irrespective of whether gravitational fields are present or not. The mathematical significance of this is that with general relativity it is not possible to find coordinates $(t;\, x, y, z)$ reducing (10.3) to (10.2) for all pairs of neighbouring events, but it is always possible to effect this reduction in a small region of space-time. This is the mathematical counterpart of what was said above, viz. that in the presence of gravitational fields frames of reference may be chosen in which special relativity is correct in any assigned neighbourhood but not over the whole of space-time.

A geometry based on (10.3) is called Riemannian. It is clear that the functional form of the $g_{\mu\nu}$ contains information about the presence and distribution of gravitational fields, but it also contains information about the particular system of coordinates chosen, which is largely arbitrary. One of the main objects of Riemannian geometry is to disentangle these two pieces of information which occur together as the form of the $g_{\mu\nu}$.

General relativity therefore takes over assumption (i) of special relativity but widens it to allow any observer, in whatever state of motion, to be used for establishing coordinates. Assumption (ii) is modified, (10.3) being the new counterpart of (10.2). (10.1) is not taken over, since the validity of Euclidean geometry (Pythagoras' theorem) is not assumed for large regions, though it is taken for granted for small ('differential') regions. The use of light signals and of standard clocks is still necessary so that intervals may be measured in the same units by different observers.

Statement (iii) is not directly applicable in a Riemannian geometry since straight lines cannot be defined unambiguously there. However, the simplest property of a straight line which can be taken over and leads, in general, to a unique result is its geodesic property, viz. that the distance between two points measured along it is a minimum (or, more generally, stationary). Accordingly, general relativity postulates that, in the absence of electromagnetic fields, but even in the presence of gravitational fields, particles follow geodesics with $ds^2 > 0$, and light rays follow geodesics with $ds = 0$. These assumptions evidently conform with the general principle of relativity that all laws of nature should be expressible in the same *form* by all observers.

The laws of general relativity as stated so far are incomplete in that they do not show by 'how much' the geometry differs from the Euclidean geometry of special relativity. To put it physically, they do not show what determines the strength of gravitational fields. Further laws are required as the relativistic counterparts of the inverse square law of Newtonian theory.

Einstein first turned his attention to the problem of how to restrict the $g_{\mu\nu}$ in empty space so as to obtain the counterpart of Laplace's equation. Of the expressions obtained in the development of Riemannian geometry the simplest one which could be used to give the required number of equations turned out to be a particular set of ten quantities (a tensor) $R_{\mu\nu}$. Each of these quantities is constructed from the 'fundamental tensor' $g_{\mu\nu}$ and its first and second derivatives with respect to the coordinates. Accordingly, Einstein chose as law of gravitation in empty space

$$R_{\mu\nu} = 0. \tag{10.4}$$

This set of second-order partial differential equations is very complicated, but Schwarzschild was able to solve it for the specially simple case in which the distribution of matter did not change with time and was spherically symmetrical and in which there was no gravitational field at infinity. This case may be considered to represent the gravitational field round the sun, and the geodesics in this geometry should correspond to the orbits of the planets and the tracks of light rays. A constant of integration occurs which represents the mass of the sun. When this has been fitted it is found that the geodesics are nearly identical with the orbits obtained in Newtonian theory. This is already highly satisfactory. A more detailed investigation shows that three of the differences between Newtonian and relativistic theory should be detectable, though only by fairly small margins. These are:

(i) A slow rotation of the axes of the elliptic orbit of the planet Mercury. Although the shift is only 43″ of arc per century it is well within observational limits, and the agreement between relativity and observation is excellent.

(ii) According to general relativity rays of light grazing the sun should be deflected by 1″·76. Newtonian theory is not very definite on this point, but predicts at most 0″·88. The observations are difficult and can be carried out only during some total eclipses, but the values obtained tend to agree with the relativistic value of 1″·76.

(iii) Spectral lines produced in a region of high gravitational potential should, according to relativity, be shifted towards the red. The ratio (change of wave-length)/(wave-length) should equal the ratio (gravitational potential)/(velocity of light)^2. Observations of solar spectral lines are difficult to interpret. Other ill-understood effects appear to mask the Einstein shift over most of the surface, but the theoretical value (2×10^{-6}) is observed on the limb. Observations have also been carried out on Sirius B, where the effect should be much larger (6×10^{-5}), but is also difficult to discern owing to the breadth of the lines.

These are the three crucial observational tests of general relativity. The motion of the perihelion of Mercury (test (i)) agrees very well with the theory, but it must be admitted that an effect of this order results from any plausible variation of Newton's law of gravitation (though some such changes given the wrong sign). The deflexion of light rays (test (ii)) is different in that it is a typical relativity effect. More observational work is greatly needed, but on the present information the test must be considered as powerful evidence in favour of the theory. Test (iii) is in a somewhat obscure state, but the balance of the evidence, ill understood as it is, certainly does not contradict general relativity.

10.3. This has completed our survey of that part of general relativity which is capable of being tested observationally (apart from cosmology). It is, however, clear that at this stage the theory is still incomplete. It has a law of gravitation for empty space (the analogue of Laplace's equation) but not for space filled with matter where an analogue of Poisson's equation is required. This is achieved in the so-called field equations

$$R_{\mu\nu} - \tfrac{1}{2} R g_{\mu\nu} = -\kappa T_{\mu\nu}. \tag{10.5}$$

In this set of ten equations the left-hand sides form a geometrical tensor, closely related to the $R_{\mu\nu}$ of (10.4), which satisfies a conservation condition (vanishing divergence) as a geometrical identity. On the right-hand side, $T_{\mu\nu}$ is a tensor constructed in a very straightforward way, which contains the density, momentum, energy and pressure of matter. κ is a constant

$$(8\pi \times \text{constant of gravitation}/(\text{velocity of light})^2)$$

which has been adjusted so as to lead to the Newtonian law of gravitation in the limiting case of small masses and densities. Equation (10.5) may be read either from left to right, showing how the presence of matter affects the geometry, or from right to left, showing how the density, momentum and energy of matter must satisfy the well-known conservation laws owing to their relation to geometrical quantities which are automatically conserved. The elegance of the field equations lies in this combination of Poisson's equation and the conservation laws of classical dynamics. Two laws of nature, previously considered to be entirely unrelated, are combined into one. On the other hand, the field equations have not been tested observationally except in so far as they lead to the Newtonian law for weak fields. Moreover, this can be achieved by substituting on the left-hand side of (10.5) any geometrical tensor which is conserved, and unfortunately there are several of these though none as simple as the one used in (10.5). Moreover, in the special case of the solar system, they would at least be compatible with all the results, enumerated above, that constitute the observational verification of the general theory of relativity. Owing to the greater simplicity of (10.5) none of these more complicated expressions is ever contemplated as a replacement of (10.5) except one, which constitutes mathematically only a very slight increase in complexity. This is

$$R_{\mu\nu} - \tfrac{1}{2}Rg_{\mu\nu} + \lambda g_{\mu\nu} = -\kappa T_{\mu\nu}. \qquad (10.6)$$

The only difference between this and (10.5) is the last term on the left-hand side, in which λ is a constant of dimensions (length)$^{-2}$. The reasons which led Einstein to introduce this, the so-called cosmological term, into his equations in a famous paper in 1916 will be discussed below in some detail. For the case of empty space, (10.6) reduces not to (10.4) but to

$$R_{\mu\nu} = \lambda g_{\mu\nu}. \qquad (10.7)$$

It is clear that no appreciable difference between the solutions of (10.4) and (10.7) will arise if λ is sufficiently small, i.e. as long as we are considering systems of dimensions small compared with $\lambda^{-\frac{1}{2}}$. The agreement between theory and observation for the solar system, arrived at from (10.4), indicates that $\lambda^{-\frac{1}{2}}$ must be very large compared with the dimensions of the solar system (at least 10^{22} cm.). This shows that the change from (10.5) to (10.6) can only be of importance in cosmological problems.

It may be of advantage to consider at this stage the position of general relativity in present-day physics. Gravitation is a phenomenon of clearly overwhelming importance for large bodies (planets, stars, etc.). For smaller bodies electromagnetic forces (classical or quantal) predominate, but as these have a self-compensating property (attraction of opposite charges) which gravitation does not share, the relative importance of gravitation increases with the size of phenomenon considered. General relativity agrees with all the observational evidence on gravitation and is in perfect accord with the local description of electromagnetic phenomena (special relativity) and the ideas belonging to it. The link does not yet extend to smaller phenomena, as the serious and obscure difficulties which stand in the way of a union of general relativity and quantum theory have not yet been overcome. General relativity is, however, the best and effectively the only description we have of the empirical world on the medium large scale, and it is natural that one should attempt to apply it on the even larger cosmological scale. It is the only theory that can provide a smooth transition between cosmology and the theory of galactic structure. On the other hand, the application to cosmology is evidently an extrapolation of an empirical theory, and may involve critical dependence on terms inappreciable locally, such as the λ term in (10.6) introduced for mathematical reasons, or the C term of Hoyle (12.8) introduced for cosmological reasons.

10.4. We now come to the curious history of the cosmological term. Einstein (1917) examined the relation of (10.5) to Mach's principle which, as he felt very strongly, should find expression in a general theory of relativity.

In his view the 'relativity of inertia' was an integral part of a 'general' theory of relativity. In other words, the inertial field defined by the $g_{\mu\nu}$ should be *fully determined* by the distribution of masses and energy in the universe. The equations (10.5) may be considered to be a set of differential equations for the $g_{\mu\nu}$ in terms of the masses, etc., represented by $T_{\mu\nu}$. The equations undoubtedly show that the $g_{\mu\nu}$ (inertia) are *affected* by $T_{\mu\nu}$ (the masses), but since (10.5) are only differential equations the $g_{\mu\nu}$ are not *determined* by them without the boundary conditions at infinity which have to be given. Now Einstein showed that it was impossible to choose boundary conditions so as to insure that the $g_{\mu\nu}$ were fully *determined* by the $T_{\mu\nu}$. Faced with this difficulty, which occurred at infinity, he introduced the cosmological term changing (10.5) into (10.6). This step had two consequences:

(i) Einstein was able to show that for positive values of λ equations (10.6) admitted of a solution in which the density of matter was uniform, its random velocities zero, and in which space was so curved that although unbounded it was finite. This meant that he abolished infinity where all his difficulties with boundary conditions arose.

(ii) He thought, though mistakenly, that for positive λ (10.6) had no solutions for $T_{\mu\nu} = 0$, that is, for empty space. Einstein was of the opinion that these two points showed that Mach's principle had been incorporated into his theory. For it is a direct consequence of the principle of the relativity of inertia that, since inertia is only determined by matter, there should be no inertia in the absence of matter. An empty universe should hence show a complete absence of inertial field so that it should be impossible to determine the $g_{\mu\nu}$ if $T_{\mu\nu} = 0$ throughout space-time. This is the significance of the demand that $T_{\mu\nu} = 0$ should have no solutions.

The system (i) is known as the Einstein universe. The astronomical data of 1917 indicated that all measured velocities of heavenly objects were small compared with the velocity of light. Accordingly, it was thought that this static system represented the universe to a first approximation. The thermodynamic difficulties that arise with a static universe have been mentioned in Chapter III but were not considered at the time. Owing to its homogeneity the Einstein universe supplies a universal standard of rest, so that it can be said that the distribution of the masses defines everywhere an inertial frame. Mach's principle is therefore satisfied in a certain sense, but no detailed method is given for showing exactly how inertia is affected by each separate mass.

Einstein's result (ii) was shown to be wrong by de Sitter in 1917. He found a solution of (10.6) for empty space ($T_{\mu\nu} = 0$). It represented an 'expanding' universe, in which test particles of negligible mass would continually recede from each other with ever-increasing velocity.

At first the main interest of 'de Sitter's universe' lay in the fact that it proved that Einstein had been wrong in his assertion (ii). Einstein has accordingly abandoned the cosmological term and with it the attempt to incorporate Mach's principle into general relativity. Before the significance of this consequence of de Sitter's discovery had been generally fully appreciated new astronomical discoveries entirely changed the climate of opinion. In the early 1920's Slipher's, Shapley's and Hubble's measurements of the radial velocities of extragalactic nebulae revealed the expansion of the universe. It became clear that the universe was not static and that astronomical objects were not always slowly moving provided only that one looked sufficiently far into space. The existence of these systematic large-scale motions not only revealed the inappropriateness of the Einstein universe, but aroused interest in the possibility of constructing a theoretical model of this 'mysterious' universe. The only one available was de Sitter's model, and observers and theoreticians grasped it as the only representation of the world which in any way fitted the stupendous discoveries that had been made. The only drawback to the identification of de Sitter's model with the actual world was then thought to be the emptiness of the model. However, it was claimed that the density of matter in the universe was in any case very low, though the meaning of this remark was not seriously discussed until much later. It was thought that there might be solutions of (10.6) intermediate between Einstein's 'matter without motion' and de Sitter's 'motion without matter', and some such solution not far from de Sitter's might well represent the actual universe. The mathematicians took up this question, and the work of Friedmann, Lemaître and Robertson solved the problem.

Before we proceed to discuss this, it may be desirable to pause to realize the effect of the developments just outlined on scientific opinion. Two conclusions of a somewhat dubious nature were drawn and their influence still persists. It was thought that the unique ability of general relativity to deal with cosmological problems had been clearly revealed, since it admitted of an expanding universe. Though this conclusion appeared to be completely upset in 1934 by Milne's and McCrea's discovery of Newtonian cosmology (described in Ch. IX, p. 123), the discussion of Layzer and McCrea in 1954 clarified and in some measure restored the position. Secondly, it has been, and still is, widely thought that the correctness of Einstein's introduction of the cosmological term into his equations had been demonstrated. The reader will have no hesitation in disregarding this opinion. There are strong arguments of a different kind for the λ term, and also against it, but Einstein's original introduction of it not only turned out to be a complete failure but showed that general relativity did not express Mach's principle fully. In the opinion of the adherents of this principle, this casts doubts on the completeness of the general theory of relativity. If a tool designed for one job fails to do it, but turns out to be useful in a different and unexpected job, one's faith in the tool is nevertheless to some extent weakened.

10.5. The first and in many ways the most fundamental discussion of the basis of relativistic cosmology was given by H. Weyl in 1923. He considered, in particular, the problem how a theory like general relativity which is based on the concept of invariance can be applied to a unique system such as the universe. General relativity was specially designed to deal with the equivalence of the observations of relatively accelerated observers. In the universe we have a single system which looks different to observers in different states of motion and which cannot be reproduced by them to suit their own convenience like a laboratory experiment. Weyl argued that in attempting to understand the distant we must base ourselves as far as possible on the theories verified in our neighbourhood. General relativity offers the best summary of all local macroscopic physics and is accordingly a suitable instrument. Further assumptions are, however, necessary in order to introduce observations not contained in local physics. The most important of these are the cosmological principle and a statement of the fact that the velocities of matter in each astronomical neighbourhood (each group of galaxies) are small. This statement was given in the form in which it has become known as Weyl's postulate: *The particles of the substratum* (*representing the nebulae*) *lie in space-time on a bundle of geodesics diverging from a point in the* (*finite or infinitely distant*) *past.*

The most important implication of this statement is that the geodesics do not intersect except at the singular point in the past (and possibly a similar singular point in the future). There is therefore one and only one geodesic passing through each point of space-time, i.e. at any point of space-time the matter passing through it possesses a unique velocity. The system of the substratum is therefore hydrodynamical, the streamlines do not intersect (except at the singular point in the distant past).

The chief significance of this assumption about the substratum is of course its relevance to the actual universe. Although the motion of the nebulae does not exactly follow this law the deviations from this general motion appear to be random and less than one-thousandth of the velocity of light. On the other hand, the relative velocities of nebulae due to the general motion is quite comparable with the velocity of light; it has been measured for some to be 30% of it and is supposed to be at least 40% of it for others that have been observed. Accordingly, the random motion may be neglected in the first instance. Combined with the observation that the motion is one of expansion rather than contraction, Weyl's postulate is seen to be closely satisfied in the actual universe.

10.6. Combining general relativity with the cosmological principle in its narrow form and with Weyl's postulate, the various models of relativistic cosmology may be derived without great difficulty. It can be shown that a bundle of geodesics satisfying Weyl's postulate must possess a set of hypersurfaces (3-spaces) orthogonal to them all. We now choose these hypersurfaces as the surfaces $t = \text{const.}$ and introduce coordinates (x^1, x^2, x^3) which are constant along the geodesics. This implies that the (x^1, x^2, x^3) of each particle is constant. Such a system of coordinates is called co-moving. The orthogonality of $t = \text{const.}$ and of $(x^1, x^2, x^3) = \text{const.}$, together with the geodesic property, is easily seen to imply that t can be chosen so that the metric is of the form

$$ds^2 = dt^2 - h_{ij}\, dx^i dx^j \quad \text{(summation in } i, j \text{ from 1 to 3 only).}$$

The h_{ij} are functions of t, $x^1 x^2 x^3$. The coordinate t plays the part of a cosmic or world time. It defines a world-wide simultaneity, and it is interesting to note that already at this stage the cosmic time has been defined, and this is essential to the ordinary (as opposed to the perfect) cosmological principle (cf. Chapter 11). Consider now a small triangle formed of three particles at some time t, and consider the triangle formed by these particles some time later. The second triangle may differ from the first in many respects. But when the postulates of homogeneity and isotropy are introduced so that no point and no direction in the 3-space may be preferential, then it follows that the second triangle must be geometrically similar to the first. Moreover, the magnification factor must be independent of the position of the triangle in the 3-space by the same postulates. It follows then that the time can enter the h_{ij} only through a common factor in order that the ratio of the distances corresponding to two small displacements may be the same at all times. Hence we must have

$$h_{ij} = [R(t)]^2 l_{ij},$$

where the l_{ij} do not depend on t. The ratio of the values of $R(t)$ at two different times is the magnification factor. The function $R(t)$ is always real,* for otherwise the lapse of time could change a space-like into a time-like interval.

Finally, we have to consider the 3-space $l_{ij}\, dx^i dx^j$, which is homogeneous, isotropic and independent of the time. Accordingly, by a well-known theorem of differential geometry, it must be a space of constant curvature. The curvature may be positive, zero or negative. With a suitable choice of x^1, x^2, x^3 and absorbing a scale factor into $R(t)$ the line element of the 4-space may then be written

$$ds^2 = dt^2 - [R(t)]^2 [(dx^1)^2 + (dx^2)^2 + (dx^3)^2]/\{1 + \tfrac{1}{4}k[(x^1)^2 + (x^2)^2 + (x^3)^2]\}^2, \quad (10.8)$$

where k has one of the values $+1$, 0 or -1. Expression (10.8) is known as the Robertson line element and was shown by him and independently by A. G. Walker to be common to all theories postulating a homogeneous isotropic substratum.

The space curvature of the 3-spaces $t = \text{const.}$ in (10.8) has the following meaning:

(i) If $k = 0$ the 3-space is Euclidean. Accordingly, the surface of a sphere of radius r is $4\pi r^2$.

* In agreement with the customary convention $R(t)$ will be taken to have the dimensions of a time. This should not cause any confusion, although the closely analogous Newtonian $R(t)$ of Chapter IX is non-dimensional. There is no direct connexion between $R(t)$ and the tensor $R_{\mu\nu}$ (or the invariant R) of equations (10.4)–(10.7).

(ii) If $k = +1$ the 3-space is spherical. This means that the surface of a sphere whose radius is of length r is *less* than $4\pi r^2$. If r is very small compared with R, then the difference is very small, but for larger r it becomes substantial until for $r = 2\pi R$ the surface of the sphere reaches a maximum. For still larger r an increase in the radius implies a diminution of the surface and for $r = 4\pi R$ the surface shrinks to a point, i.e. any geodesic of this length returns to the starting-point. In general, the surface of a sphere is $64\pi R^2 \sin^2 (r/4R)$. The space is closed and has volume $\pi^2 R^3$.

(iii) If $k = -1$ the 3-space is hyperbolic. This means that the surface of sphere of radius r is *greater than* $4\pi r^2$. For r small compared with R, the difference is negligible, but for large r it becomes very large. In general the surface is $64\pi R^2 \sinh^2 (r/4R)$. The space is open.

The Robertson line element (10.8) expresses the consequences of the cosmological principle and of Weyl's postulate for the geometry of space. We must now investigate its dynamical significance by using the field equations (10.6). It is easily seen that the only surviving terms in $T_{\mu\nu}$ correspond to a material density $\rho(t)$ and an isotropic pressure $p(t)$, both of which are functions of the time only. The explicit relations are

$$\kappa\rho = -\lambda + 3\frac{k+\dot{R}^2}{c^2R^2}, \quad \frac{\kappa p}{c^2} = \lambda - \frac{2R\ddot{R}+\dot{R}^2+k}{c^2R^2}, \qquad (10.9)$$

where a dot denotes differentiation with respect to the time. These equations show what values of the density and of the pressure correspond to any given set of the constants λ and k and the function $R(t)$. Physical significance can only be assigned to a set which gives both $\rho > 0$ and $p \geqslant 0$. It should be observed that p includes all types of pressure, such as that due to the random motions of the nebulae and stars, and that due to the heat motion of molecules, radiation pressure, etc.

An interesting relation may be constructed by considering a volume bounded by a set of fundamental particles. Let the volume enclosed be V. Then clearly, owing to the motion of the substratum, $V \sim R^3(t)$. If we now call the total mass (energy) within the volume $E = \rho V$, then it is easily verified that as a consequence of (10.9)

$$dE + p\,dV = 0. \qquad (10.10)$$

This is the law of conservation of energy and shows that the pressures do work in the expansion. The Newtonian equations (Chapter IX) differ from this, since there the strict conservation of mass alone leads to

$$dE = 0, \qquad (10.11)$$

which is equivalent to (9.16).

In the actual universe, observations show that the pressure (i.e. the energy density of all types of random motion) is far smaller than the energy density due to matter. The ratio of the two quantities is about 10^{-5} or 10^{-6}. Accordingly, as long as only states of the universe differing not too widely from the present one are contemplated, p may be put equal to zero in the second of equations (10.9) or alternatively in (10.10). Then

$$R(\dot{R}^2 + k) - \frac{c^2}{3}\lambda R^3 = \tfrac{1}{3}\kappa R^3 \rho = \text{const.} = C \quad (\text{say}), \qquad (10.12)$$

so that

$$\left(\frac{dR}{dt}\right)^2 = \frac{C}{R} - k + \frac{c^2}{3}\lambda R^2. \qquad (10.13)$$

This relativistic equation for the variation of the scale factor R with t in the absence of pressures is identical with the corresponding Newtonian equation (9.16). The interpretation of the terms is also identical except that the constant k represents the curvature of the 3-space orthogonal to the world lines of the fundamental particles in general relativity, whereas it represents the total energy of the particles in Newtonian theory. This geometrical representation of Newtonian total energy is not confined to cosmological models in general relativity. But for this fact, which is of considerable importance in the interpretation of the theory, the identity of the relativistic and the Newtonian equation is complete.

The consequences of this remarkable similarity discovered by Milne and McCrea in 1934 are of great significance. In the first instance it shows that the difference between relativistic and Newtonian theories is governed, in cosmology, by the ratio of pressure to density, and that the ratio of gravitational potential to rest mass is important mainly in local applications. Hence in a universe like ours is now, where both these ratios are always very small, relativity cannot offer anything radically new. It has the great advantage over Newtonian theory of giving definite answers to questions concerning the propagation of light. It will give somewhat different results for light rays travelling over distances comparable with the radius of curvature of the universe. But it cannot be expected to give any radically non-Newtonian results for 'intermediate size' problems such as the spiral structure, the clustering, or the condensation of the nebulae.

Secondly, it shows that the ability of general relativity to deal with cosmological questions is not in as great contrast to Newtonian theory as was once thought. For problems in which neither the pressure/density nor the (gravitational potential)/(rest mass) ratios are appreciable, relativity gives answers nearly the same as Newtonian theory, though they are more easily and definitely interpreted. This advantage is frequently obscured by the very great (often insoluble) mathematical difficulties encountered in the relativistic treatment.

Thirdly, it shows that the history of the universe as described by the variation of the scale factor $R(t)$ is identical for the two theories. Accordingly, there is no need to repeat here the discussion of the various models given in Chapter IX.

It is only necessary to mention in addition the de Sitter universe which is that solution of (10.19) for which $k = 0$ and $\dot{R}/R$ is constant. Then $\ddot{R}/R = (\dot{R}/R)^2$, so that $\rho = -p/c^2$. The only possibility that gives neither a negative density nor a negative pressure is

$$\rho = p = 0, \quad R(t) \sim \exp\,[t(\lambda/3)^{\frac{1}{2}}].$$

The constant λ must be positive. This model has the remarkable property that it is stationary, i.e. it presents the same aspect at all times. The metric is invariant against a shift of t and a suitable simultaneous change of scale of the space coordinates.

In spite of this interesting property the model is of little direct interest in relativistic cosmology since it is void of matter. It is, however, the common limiting case to which all indefinitely expanding models with positive λ tend as $t \to \infty$.

The discussion of the propagation of light is quite simple in general relativity. By changing from Cartesians to spherical polars and then appropriately altering the radial coordinate,* the Robertson line element (10.8) can be put into the form

$$ds^2 = dt^2 - R^2(t)\left[\frac{dr^2}{1-kr^2} + r^2(d\theta^2 + \sin^2\theta\, d\phi^2)\right]. \quad (10.14)$$

If the coordinates are chosen so that the observer is at the origin $r = 0$, it is easily seen from the spherical symmetry of the system that for any radial ray of light both θ and ϕ must be constant. By the postulates of the theory $ds = 0$ along a light ray, and therefore

$$\frac{dt}{R(t)} = \pm\frac{dr}{(1-kr^2)^{\frac{1}{2}}}. \quad (10.15)$$

Hence

$$\int_{t_1}^{t_2} \frac{dt}{R(t)} = \int_0^{r_1} \frac{dr}{(1-kr^2)^{\frac{1}{2}}} \quad (10.16)$$

gives the connexion between the times t_1 and radial coordinates r_1 of events observed at the origin at time t_2.

For two light rays emitted by the same fundamental particle (same r_1) at t_1 and a short time dt_1 later, (10.16) gives for the difference dt_2 of the times of observation

$$\frac{dt_2}{R(t_2)} = \frac{dt_1}{R(t_1)}, \quad (10.17)$$

since the right-hand side of (10.16) is unaltered. Now for a fundamental particle (r, θ, ϕ constant) $dt = ds$, so that dt_1 and dt_2 are the proper time intervals between the rays, measured at the source and observer respectively. Accordingly, if the two light rays are sent out by an atom at r_1 corresponding to the peaks of the oscillation corresponding to a spectral line, that interval will appear to O to be multiplied by $R(t_2)/R(t_1)$, i.e. the frequency is divided by this factor. The observer will therefore notice a Doppler shift

$$1 + z = \frac{\nu_1}{\nu_2} = \frac{dt_2}{dt_1} = \frac{R(t_2)}{R(t_1)}. \quad (10.18)$$

This again is identical with the corresponding Newtonian formula.

* In general relativity the coordinate is only a label, and is connected with distance only by the metric. The r-coordinate used here is purely conventional and should not be confused with the radial distance r used in Chapter IX.

If, roughly speaking, source and observer are near to each other, (10.18) may be written

$$1+z = \frac{R(t_2)}{R(t_2-\Delta t)} = \frac{R(t_2)}{R(t_2)-\Delta t\dot{R}(t_2)} = 1+\Delta t\frac{\dot{R}(t_2)}{R(t_2)}. \quad (10.19)$$

Equation (10.16) can then be integrated approximately to give

$$\frac{\Delta t}{R(t_2)} = r_1, \quad (10.20)$$

so that

$$z = r_1\dot{R}(t_2). \quad (10.21)$$

Accordingly, the Doppler shift is proportional to r_1, so that we can speak of a velocity-distance relation (cf. Chapter v).

To make these statements more precise, a clear definition of distance is required. Mathematically, it is easy to define an 'absolute' distance between two fundamental particles by considering them at the same value of t. It is then clear from (10.14) with $dt = 0$, $d\theta = d\phi = 0$, that

$$D_1 = R(t)\int_0^{r_1} \frac{dr}{(1-kr^2)^{\frac{1}{2}}} \quad (10.22)$$

gives a definition of the distance of the event from O at time t.

Physically this definition is, however, not very valuable. In practice all measurements of distances of nebulae are made by measuring the apparent luminosity of the nebula or one of its stars and comparing it with the absolute luminosity of the object which is inferred in the way discussed in Chapter v. We must therefore examine how the energy L_1 sent out by the source E per unit time (both energy and time being measured in E's proper units) compares with l_2, the energy per unit area per unit time received by O, all quantities there being measured in O's units. The problem is easily solved by using the tensor calculus and the result is

$$l_2 = \frac{L_1 R^2(t_1)}{4\pi r_1^2 R^4(t_2)\,c^2} = \frac{L_1}{4\pi r_1^2 R^2(t_2)\,(1+z)^2\,c^2}. \quad (10.23)$$

(The Newtonian formula (9.14) is identical with this.)

This rather complicated looking formula can be interpreted quite easily. Consider, in the 3-space, the area of the surface of the sphere with centre at E and passing through O. It equals, owing to the homogeneity of the 3-space, the surface area of the sphere with centre at O passing through E. *At t_2, i.e. at the moment of observation*, this area is $4\pi R^2(t_2)\, r_1^2$. Formula (10.23) shows that the power has to be divided not only by this area, as is evident in the classical picture, but also by $R^2(t_2)/R^2(t_1)$, that is, by the square of the Doppler shift. This division by the square of the Doppler shift was ill understood at one time, but Roberston cleared up the question entirely in 1937 and firmly established (10.19). That *one* division by the Doppler shift is necessary is immediately clear, since the interval of time during which a certain burst of energy is received is longer than the interval of emission by virtue of the Doppler shift. Any measurement of power must be corrected for the different rates of passage of time at source and observer by just this factor. The second factor arises since, in general relativity, energy is not an invariant but the time component of a four-vector. Hence the energy emitted by the source is measured differently by an observer moving with the source and by an observer elsewhere. The second factor takes account of this difference in energy measurement.

Accordingly, the luminosity distance, as it is called, of an object is defined as the root of the ratio of rate of emission of energy of 4π times the rate of reception of energy per unit area, corrected by the square of the Doppler shift. Hence

$$\text{luminosity distance } q_1 = [L_1/4\pi l_2(1+z)^2]^{\frac{1}{2}}. \qquad (10.24)$$

By formula (10.19)
$$q_1 = cr_1 R(t_2). \qquad (10.21')$$

We can now give a more precise interpretation of the velocity-distance relation (10.21): The ratio of the Doppler shift to the luminosity distance approaches the ratio $\dot{R}(t_2)/R(t_2)$ when a source sufficiently near the observer is chosen. For objects too far away the higher-order terms in (10.20) can no longer be neglected, but in practice their contribution seems to be inappreciable for nebulae brighter than the 18th magnitude, which includes all the nebulae whose Doppler shifts have so far been measured. For objects too close to the observer the random motions which have been excluded from our model do in fact obscure the general motion, but there is, as has been discussed in Chapter V, a good range of nebulae satisfying the velocity-distance law. They give a good determination of Hubble's constant which in this theory must be identified with $\dot{R}(t_2)/R(t_2)$.

The exact connexion between the observable quantities z and q follows from (10.16), (10.18) and (10.24) if r and $R(t)$ are eliminated. A third observable quantity is the number of nebulae $N(m)$ of apparent magnitude greater than m. In order to evaluate this quantity in the theory it is necessary to introduce $n(t)$, the number of nebulae per unit volume. This will be a function of t only. Consider now the 3-space corresponding to the instant of observation $t = t_2$, i.e. the world map, not the world picture of the observer O. The volume enclosed by the sphere with O at the centre and bounded by $r = r_1$ is

$$V = 4\pi R^3(t_2) \int_0^{r_1} \frac{r^2\, dr}{(1-kr^2)^{\frac{1}{2}}}. \tag{10.25}$$

The number of nebulae now in this volume is

$$N = Vn(t_2). \tag{10.26}$$

In relativistic cosmology the assumption is usually made that all nebulae existing at $t = t_2$ have existed also at the time of emission of light $t = t_1$. Since for the most distant objects yet observed $t_2 - t_1 \doteqdot 5 \times 10^9$ years, the correctness of this assumption is somewhat in doubt. In order to replace it by something else far more would have to be known about nebular evolution than is known now. Mathematically, this assumption amounts to putting $R^3(t)\, n(t) =$ constant. Accordingly, (10.26) is supposed to represent the number of nebulae in the sphere $r \leqslant r_1$ at all relevant times.

10.7. Equations (10.16), (10.18), (10.21) and (10.23) establish the connexion between the observable quantities. The detailed comparison of theory and observation forms a most difficult and intricate subject. Its chief purpose, at this stage, is to discover which of the relativistic models (if any) fit the actual universe.

A brief outline of the subject will be given here, but in the form appropriate to the 1950 data, as no full re-discussion with modern results is available. Hence this section and the next one are chiefly of historical interest.

Theory generally deals with the *total* energy received per unit time from each luminous object, but observations are generally carried out by means of photographic plates sensitive to a comparatively small range of the spectrum only. The total intensity is expressed as apparent bolometric magnitude

$$m_b = \text{const.} - 2{\cdot}5 \log_{10} l_2.$$

Then by (10.19)

$$m_b = -5 \log_{10} [r(1+z)] + \text{const.}, \tag{10.27}$$

where the constant contains the units of m_b, $R(t_2)$ (which is the same for all nebulae since all observations are carried out at effectively the same time) and L_1. The assumption currently made in relativistic cosmology is that L_1 is constant. This implies not only that the different individual luminosities of the nebulae are averaged out, but that the average absolute luminosity is effectively independent of the time. The constancy of L_1 and of nR^3 form the two assumptions on which the comparison of theory and observation rests. Their validity can be relied on only for comparatively near objects whose light was emitted not too long ago. But for these objects the typical relativity effects such as space curvature are too small to be evaluated from the observations with their inaccuracies. Looking farther and farther into space is almost valueless, since any increase in definiteness of information is matched by a decrease in the reliability of the assumptions $nR^3 = \text{const.}$, $L_1 = \text{const.}$

Changing from bolometric to photographic magnitudes m_p introduces the K-term (cf. Chapter V). Measuring $R(t_2)$ ($=R_2$ for brevity) in parsecs we have

$$m_p = M_p - K(z) + 5 \log_{10} [r(1+z)] + 5 \log_{10} R_2 - 5. \tag{10.28}$$

M_p is the absolute magnitude of the nebula defined as its apparent magnitude if it were moved to a position only 10 parsecs from us. From (10.16)

$$\int_{t_1}^{t_2} \frac{dt}{R(t)} = \begin{cases} \sin^{-1} r & (k = +1), \\ r & (k = 0), \\ \sinh^{-1} r & (k = -1). \end{cases} \tag{10.29}$$

Also

$$1 + z = R_2/R_1. \tag{10.30}$$

Finally, from (10.26) and writing A for the constant $4\pi R_2^3 n(t_2)/3$,

$$\frac{N}{A} = \begin{cases} \frac{3}{2}[\sin^{-1} r - r(1-r^2)^{\frac{1}{2}}] & (k=+1), \\ r^3 & (k=\ \ 0), \\ \frac{3}{2}[r(1+r^2)^{\frac{1}{2}} - \sinh^{-1} r] & (k=-1). \end{cases} \tag{10.31}$$

Various methods of dealing with this set of complex equations have been devised. The quantities m_p, z and N are observable, whereas r and t are only of theoretical significance. The elimination of r and t is very difficult. Power-series solutions are of dubious validity for application to the most distant nebulae observed, but they illustrate very well the orders of magnitude of the various effects. We shall confine ourselves to a description of this method. The more accurate methods devised mainly by Heckmann are given in his book; the treatment given here largely follows McVittie (1939).

It seems likely from all investigations of the K-term that it can be expanded as a power series in z for small z. Accordingly, we put

$$K(z) = 5\log_{10}(1+az+bz^2+\ldots). \tag{10.32}$$

Then (dropping the suffix p since all magnitudes referred to henceforth will be photographic)

$$\begin{aligned} 0{\cdot}2(m-M) &= \log_{10}[r(1+z)(1+az+bz^2+\ldots)] + \log_{10} R_2 - 1 \\ &= \log_{10} r + \log_{10} R_2 - 1 + \frac{1}{E}[(1+a)z + (a+b)z^2 + \ldots], \end{aligned} \tag{10.33}$$

where $1/E = \log_{10} e = 0{\cdot}4343$, $E = 2{\cdot}3026$.

Also by (10.31)

$$\log_{10} N = \text{const.} + 3\log_{10} r + \frac{3}{10E} kr^2 + \ldots. \tag{10.34}$$

We now put $t_2 - t_1 = \Delta$ and expand $R(t)$ in (10.29) and (10.30). Then

$$\frac{\Delta}{R_2} + \frac{\Delta^2}{2}\frac{\dot{R}_2}{R_2^2} + \ldots = r + \tfrac{1}{6}kr^3 + \ldots, \tag{10.35}$$

whereas

$$z = \Delta\frac{\dot{R}_2}{R_2} + \Delta^2\left(\frac{\dot{R}_2^2}{R_2^2} - \frac{\ddot{R}_2}{2R_2}\right) + \ldots. \tag{10.36}$$

From now on the suffix 2 can be omitted, all values referring to the time of observation.

Consider now the expression $0{\cdot}6m - \log_{10} N$. In a homogeneous static Euclidean universe it would be constant. In the relativistic model, however,

$$0{\cdot}6m - \log_{10} N = \text{const.} + \frac{3}{E}[(1+a)z + (b - \tfrac{1}{2} - \tfrac{1}{2}a^2)z^2 - \tfrac{1}{10}kr^2] + \ldots. \tag{10.37}$$

We now eliminate Δ between (10.35) and (10.36). Then

$$r = \frac{z}{\dot{R}} - \frac{z^2}{2\dot{R}}\left(1 - \frac{R\ddot{R}}{\dot{R}^2}\right) + \ldots. \tag{10.38}$$

Returning to (10.33), which is now only required to order z,

$$0{\cdot}2m = \log_{10} z + \frac{z}{E}\left(\frac{1}{2} + a + \frac{1}{2}\frac{R\ddot{R}}{\dot{R}^2}\right) + \ldots + X, \tag{10.39}$$

where
$$X = 0{\cdot}2M + \log(R/\dot{R}) - 1. \tag{10.40}$$

Then
$$10^{0{\cdot}2m} = 10^X z\left[1 + z\left(a + \frac{1}{2} + \frac{1}{2}\frac{R\ddot{R}}{\dot{R}^2}\right) + \ldots\right]. \tag{10.41}$$

Let
$$u = 10^{0{\cdot}2m}, \quad v = \frac{E}{3}[0{\cdot}6m - \log_{10} N].$$

Then by (10.37), (10.38) and (10.41)

$$v = \alpha_0 + \alpha_1 u + \alpha_2 u^2 + \ldots, \tag{10.42}$$

where

$$\alpha_1 = (1+a)\,10^{-X}, \tag{10.43}$$

$$\alpha_2 = \left[a + b - (1+a)\left(a + \frac{1}{2} + \frac{1}{2}\frac{R\ddot{R}}{\dot{R}^2}\right) - \tfrac{1}{2}(1+a)^2 - \frac{k}{10\dot{R}^2}\right]10^{-2X}$$

$$= \left[-1 - \tfrac{3}{2}a - \tfrac{3}{2}a^2 + b - \tfrac{1}{2}(1+a)\frac{R\ddot{R}}{\dot{R}^2} - \frac{k}{10\dot{R}^2}\right]10^{-2X}. \tag{10.44}$$

The evidence (now partly superseded) to which these formulae could be applied in 1950 was as follows:

(i) The observations of the velocity-distance relation (p. 38).

(ii) Observational estimates of the average absolute magnitude of nebulae (p. 37).

(iii) The theory of the K-term (p. 49).

(iv) Observational estimates of the effective temperatures of the nebulae (p. 38).

(v) Observational estimates of ρ, the mean density of matter, and p, the average pressure (p. 45).

(vi) The relativistic formulae (10.8) connecting p, ρ, λ, k and $R(t)$.

(vii) The nebular counts (p. 41).

The observations dealing with the velocity-distance relation serve, in the first instance, to determine the constant X in equation (10.39). Some authors also use these observations to determine the coefficient of z in (10.39), but in view of the uncertainties of the observations this is of questionable value, and we shall content ourselves with deducing that this coefficient cannot be very large. The value of X is probably between 4·7 and 5·0 (p. 40).

Combining this result with estimates of the average absolute magnitudes of nebulae the constant $T = R/\dot{R}$ may be determined. Its value is found to be about $1{\cdot}8 \times 10^9$ years $= 5{\cdot}5 \times 10^{16}$ sec. The uncertainty of T is not as great as might be expected from the uncertainty of X, since the higher estimate of X corresponds with the higher estimate of M.

The theory of the K-term shows that for a narrow band of sensitivity and black-body radiation

$$b = 1 + \frac{a^2}{2}. \tag{10.45}$$

For a wider band of sensitivity b is a little smaller. We put

$$b = 1 + \frac{a^2}{2} - \sigma, \tag{10.46}$$

where the positive quantity σ is almost certainly less than 0·5 for the sensitivity curve of the photographic plates employed.

Eliminating λ between the relativistic formulae (10.9) and making use of the observational fact that p effectively vanishes, we find

$$4\pi\gamma\rho T^2 = k\frac{c^2T^2}{R^2} + 1 - \frac{R\ddot{R}}{\dot{R}^2}. \tag{10.47}$$

We can now rewrite formula (10.44) in the form

$$\left[1 + \frac{1}{5\alpha_1 10^X}\right] k\frac{c^2T^2}{R^2} = 4\pi\gamma\rho T^2 - \Psi, \tag{10.48}$$

where $$\Psi = \frac{2}{\alpha_1 10^X}[(\alpha_1^2 + \alpha_2)\, 10^X - \tfrac{1}{2} - \sigma].$$

The number counts may now be used to find likely values for α_1 and α_2. The value of α_1 helps to determine a and X, which are already quite well known from the work on the K-term and on the velocity-distance relation. The parameters α_1 and α_2 together lead by (10.48) to a relation between the curvature of space and the density of matter. Earlier views that the number counts could determine both the curvature and the density were shown to be incorrect by McVittie (1939).

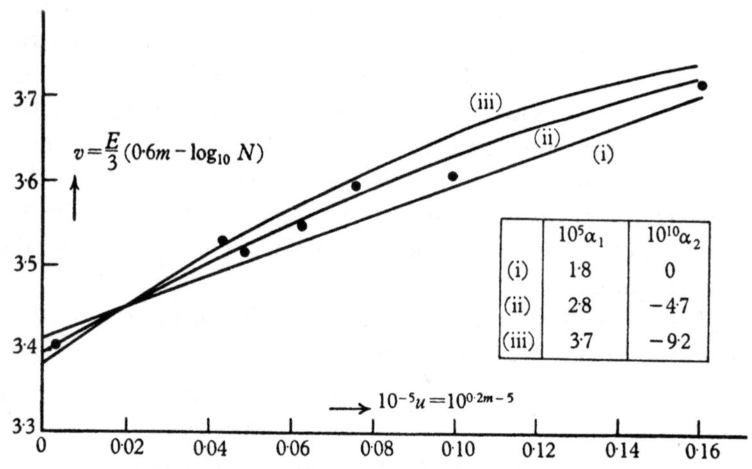

Fig. 5. Comparison between relativistic cosmology and observation.
• Nebular counts (Table I, p. 41).
— Theoretical curves (Table III, p. 115).

It can be seen readily that the number counts are best represented with

$$10^5\alpha_1 = 2{\cdot}8, \quad 10^{10}\alpha_2 = -4{\cdot}7$$

and that

$$10^5\alpha_1 = 1{\cdot}8, \quad 10^{10}\alpha_2 = 0 \quad \text{and} \quad 10^5\alpha_1 = 3{\cdot}7, \quad 10^{10}\alpha_2 = -9{\cdot}2$$

give fairly extreme interpretations of the results (see Fig. 5).

Formula (10.41) for α_1 may now be used, in conjunction with a knowledge of X, to deduce the value of a and hence of the effective temperature of nebular radiation.

Table III lists these, as well as Ψ, the last term on the right-hand side of (10.48), for the extreme cases $X = 4{\cdot}7$, $X = 5$ and $\sigma = 0$, $\sigma = \frac{1}{2}$.

The significance of the figures in the last two rows is as follows: Any definite choice of α_1, α_2, X and σ leads to one value of Ψ; if $4\pi\gamma\rho T^2$ has just this value, then space is flat; if $4\pi\gamma\rho T^2$ exceeds this value then k is positive, whereas if it is less than this value then k is negative; since the factor of kc^2T^2/R^2 is always between 1·05 and 1·25 the magnitude of the curvature is almost equal to the difference between $4\pi\gamma\rho T^2$ and Ψ.

Table III

$10^5\alpha_1$	1·8		2·8		3·7	
$10^{10}\alpha_2$	0		−4·7		−9·2	
X	4·7	5	4·7	5	4·7	5
a	−0·1	0·8	0·4	1·8	0·85	2·7
Effective temp.	∞	6000°	7500°	4500°	5700°	3500°
Ψ $\sigma = 0$	0·7	3·1	0·5	1·9	0·6	2·2
Ψ $\sigma = \frac{1}{2}$	−0·4	2·5	−0·3	1·7	0·1	1·9

Whether any further conclusions can be drawn depends on the weight one assigns to various observations and what further assumptions one is prepared to make. Hubble (1936) took $X = 4{\cdot}7$ and $a = 0{\cdot}80$ to be very well determined results, and accordingly the solution of the penultimate column is the only one close to his solution. He also assumed implicitly that $R\ddot{R}/\dot{R}^2$ was close to unity and so was led to a solution with high density ($\rho = 6 \times 10^{-27}$ g./cm.^{3}) and large positive curvature ($Rk^{-\frac{1}{2}} = 3 \times 10^8$ light-years).

As he pointed out, this universe has a radius not much larger than the distance of the most distant nebulae observed and a density much larger than he expected. He found this result very unsatisfactory, partly because of a philosophical objection to a universe which has been largely explored and partly because of the current prejudice against high densities (this was discussed in Chapter v). He showed that both these features, in his view objectionable, disappeared if formula (10.23) was amended so that the Doppler shift appeared only singly and not through its square. He gave no theoretical reasons for this, though he presumed that if the nebulae were not 'truly receding' and the red shift was due to some unknown cause, then the amended formula (10.23) would hold. In fact, any such assumption is incompatible with general relativity though it finds expression in kinematic relativity (Chapter xi).

McVittie (1939) made the inaccurate assumption that $b = 0$. He found that unless ρ was much higher than his rather low estimate (2×10^{-29} g./cm.^{3}) the curvature was negative and the radius of curvature rather large ($2{\cdot}5 \times 10^{10}$ light-years).

Heckmann (1942), using more recondite methods, lends a little more support to the Hubble-Tolman conclusion that k is positive, but shows how sensitive this result is to very small observational errors.

We may sum up this discussion by saying that the pre-1940 astronomical observations did not point clearly to any one of the relativistic models nor can they be construed to disagree seriously with general relativity.

Whether observations reaching farther out into space can lead to more definite conclusions depends on the correctness of the assumptions nR^3 = const., L = const. for periods of time greater than 5×10^8 years. Without very great progress in the theories of stellar and nebular evolution little of definite value for relativistic cosmology can therefore be expected to follow from the progress of astronomical techniques alone.

10.8. So far the discussion has been confined to a comparison of the theory and astronomical observations of distant nebulae. Far greater difficulties arise when the theory is compared with the other data available. In particular, the estimates of the time-scale of the universe arrived at from other sources lead to very important conclusions.

It will be remembered that the age of the Earth has been shown to be at least 3×10^9 years, and many stars seem to have existed for about 5×10^9 years. If the universe has existed for a finite time only, then its age must be at least 5×10^9 years, even allowing for an appreciable margin of error. Before 1952, the observations were believed to show that $R_2/\dot{R}_2 \doteqdot 1{\cdot}8 \times 10^9$ years. Comparison with the age estimates thus led to the conclusions described below, now largely superseded.

Consider the graph of $R(t)$ against t. The tangent to the curve at $t = t_2$ intersects the line $R = 0$ at a point t_3. Since the universe is expanding $t_3 < t_2$. Also $t_2 - t_3 = R_2/\dot{R}_2 = 1{\cdot}8 \times 10^9$ years. Now if $\ddot{R} \leqslant 0$ for all $t \leqslant t_2$, then the model started off from a point at a time less than $1{\cdot}8 \times 10^9$ years ago. For then the steepness is constantly decreasing so that the entire curve must lie to the right of the tangent at $t = t_2$. Any such model must therefore be rejected, since it disagrees with the observations referred to.

Now it follows from (10.9) that

$$\kappa\left(\rho + 3\frac{p}{c^2}\right) = 2\lambda - 6\ddot{R}/R.$$

Since the left-hand side is always positive $\ddot{R}/R$ will always be negative unless $\lambda > 0$. Therefore λ must be positive for any relativistic model of the universe that attempts to agree with current determinations both of Hubble's constant and of the age of the galaxy. A few authors are of the opinion that general relativity is so well founded and that these determinations are so uncertain that the case of $\lambda = 0$, which is possibly more desirable theoretically, should not be rejected at this stage. However, most of the work on relativistic cosmology is based on these observational results and hence assumes that $\lambda > 0$. The possibilities for the development of the universe are then easily picked out from the solutions described in Chapter IX. Using the system of classification employed there, essentially three different cases must be considered:

A. *Cases* 1 (i), 2 (i), 3 (i). The model has a point origin and expands indefinitely. There is a value of R at which the rate of expansion is a minimum. The only difference between these models is the value of k.

B. *Case* 3 (ii *c*). This model starts from a finite size ($R = R_c$) and expands indefinitely with ever-increasing speed.

C. *Case* 3 (iii *b*). The model contracts from infinity with ever-diminishing speed to a finite minimum size and then expands with constantly increasing speed.

Model 3 (ii *a*) (Einstein's universe) must be rejected since there is no expansion. Models 3 (ii *b*) and 3 (iii *a*) must be rejected since the expansion is slowing down.

For a considerable period Model 3 (ii *c*) was the most favoured. It had been discovered by Lemaître in 1925, and Eddington argued powerfully in its favour. It is best pictured by reference to the static Einstein universe in which gravitational attraction and 'cosmological' (i.e. λ) repulsion are just in equilibrium. As Eddington pointed out this model is unstable. For if, for some reason, it should expand very slightly then this would diminish the gravitational forces holding it together, but increase the repulsion since gravitation diminishes but λ repulsion increases with increasing separation of the objects. Accordingly, the model would continue to expand and the speed of expansion constantly increase as the model moves farther and farther away from the equilibrium state. Similarly, a small initial contraction of the Einstein universe will lead to further contraction, corresponding to Model 3 (ii *b*) in reverse.

The Lemaître-Eddington model has therefore an infinite past which was spent in the Einstein state. This has greatly attracted investigators since it seemingly permits an arbitrarily long time-scale of evolution. The picture of the history of the universe derived from this model, then, was that for an infinite period in the distant past there was a completely homogeneous distribution of matter in equilibrium in the Einstein state until some event started off the expansion, which has been going on at an increasing pace ever since. The condensation of the galaxies and the stars from the primeval matter took place at the time the expansion began, but this development was stopped later by the decrease of average density due to the progress of the expansion.

Attention was soon focused on the two closely allied questions of what the nature of the event was that pushed the system out of equilibrium, and why the resultant motion was one of expansion rather than contraction. It was clear that it was desirable to establish a linkage between the expansion of the universe and the condensation of the nebulae. The primeval matter was pictured as a uniform gas at low (possibly zero) temperature, and such material is unstable against gravitational condensations. Could this local instability possibly be merely a more familiar aspect of that newly discovered phenomenon, the instability of the Einstein state? The problem is not at all simple owing to the difficulties of the mathematics and of interpretation in general relativity. The work of Lemaître, McCrea and McVittie did not settle the question completely but suggested rather strongly that the guess was correct, that the formation of a local condensation would make an Einstein universe expand.

This answer seemed to complete the round of arguments in favour of the Lemaître-Eddington model. But since then the wide, though not general, tendency has been to regard the model as unsatisfactory. A number of questions can be asked which are at best awkward to answer on the basis of this model.

If one particular condensation started off the whole process of expansion one would expect that condensation and its neighbourhood to differ from the rest of the universe. No such distinctive condensation is known from observation, and its existence would be incompatible with the cosmological principle.

If the atomic structure of matter is considered, the existence of the Einstein state becomes quite inexplicable. For since any condensation leads to the onset of the instability, the atoms must be very smoothly distributed. This must imply a regular crystal structure, and such a structure cannot exist in a curved spherical space if there are more than a certain low number of atoms. Since the number of atoms is certainly quite large it follows that the existence of the Einstein state is incompatible with the atomic structure of matter.

Another question concerns the formation of galactic condensations. The occurrence of such a condensation must have been extremely unlikely at the temperature and density of the Einstein state, for otherwise the life of that state would be very short. But once the expansion starts circumstances become still less favourable owing to the decrease in density. Hence it is difficult to see how so very many galaxies have come to be formed after the first one initiated the expansion process.

Unfortunately, these and similar objections have hardly been discussed and replied to, but, for very different reasons, there came about 1935 a swing of opinion away from the Lemaître–Eddington model.

The question of the generation of the elements has already been mentioned (Chapters I and VII). It is considered desirable by many authors to show how the heavy elements have been generated from hydrogen, the simplest and most primitive element. Modern nuclear physics appeared to show that the building up of heavy elements could only take place in conditions of extremely high density and temperature. The point-source models of general relativity appeared to offer a possibility of a place and time at which such extreme conditions have occurred. For as $R(t) \to 0$, the density tends to infinity as R^{-3}, and the temperature as R^{-1}. The purely gravitational and inertial terms will not be dominant in such conditions, and accordingly we cannot follow the model up to its origin using merely the field equations (10.9). However, it is clear that the point-source models must have had very high densities and temperatures, even though the details of the development cannot be examined so easily. This possibility of linking up the generation of the elements with the origin of the universe has been examined by Gamow and Teller (1939), Alpher and Herman (1950), and others, mainly from the nuclear physics point of view. These authors have attempted to deduce the conditions prevailing at the time of the origin of the universe from the present frequencies of occurrences of the heavier elements. They have not specified too closely the type of point-source model, but this can be deduced to some extent from the evolution of the nebulae. Lemaître has investigated this problem, and in his view three periods in the evolution of the universe should be distinguished. At first there was a period of explosion from a point source, during which the elements were formed. The velocities of expansion were too high at that time for condensations to be formed. The next period was one of very much reduced rate of expansion in which the large-scale motions were almost insignificant, but during which conditions were very favourable for the formation of condensations. The enormous local deviations from homogeneity that occur in the universe are ascribed to events taking place during this second period. The third period is one of renewed expansion. The recession of the nebulae is accelerating and the formation of new condensations is made unlikely by the diminishing density.

The mathematical formulation of this three-period development is contained, as Lemaître showed, in the relativistic model 3 (i). The three models 1 (i), 2 (i) and 3 (i) each have a period of decreasing speed of expansion, but only in 3 (i) can the expansion be brought almost to a standstill. For if $\lambda = \lambda_c$ we are dealing with 3 (ii*b*) in which the expansion tends to zero. If λ very slightly exceeds λ_c then 3 (i) consists therefore of a development similar to that of 3 (ii*b*), but gravitation can never completely counteract the cosmic repulsion, whereas if $\lambda = \lambda_c$ the two are asymptotically equal. Hence, in the model in question, the Einstein state is not reacned but only, as it were, touched and the development of 3 (ii *c*), the old Lemaître-Eddington model, follows.

Lemaître's model (3 (i)), with λ slightly greater than λ_c, has many attractive features and, especially if combined with the work of Gamow and Teller, seems to be the best relativistic cosmology can offer. The time-scale difficulty is largely resolved through the interposition of the arbitrarily long 'quasi-Einstein' stage. The occurrence of the heavy elements was thought to be due to the initial conditions of temperature and density (see p. 57).

The theory makes two definite statements which it may soon be found possible to check against observation. It can easily be shown that according to this theory the product of the constant of gravitation and the average density divided by the square of Hubble's constant (which, as was pointed out in Chapter VII, is a pure number) should be less than about $\frac{1}{6}$, but probably not very much less. This means that the average density should be less than about $1{\cdot}5 \times 10^{-29}$ g./cm.^{3}. As was discussed in Chapter V there are some schools of thought according to which present observations indicate a somewhat higher density, though most observational astronomers would consider a lower value more likely.

A second point is common to all relativistic theories in which λ is positive. Since, according to these theories, the present expansion indicates that cosmic repulsion is, and for some time has been, more powerful than the gravitational action of the smoothed-out universe, the condensation of new galaxies from the general background is no longer possible. It has been proved by various authors that the cosmic repulsion does not destroy gravitationally bound systems of high average density like the solar system or the galaxies, but there can be little doubt that it precludes the formation of new condensations from material of low density. Any such theory therefore implies that the ages of the nebulae have a fairly definite lower limit, i.e. hardly any nebula can be younger than about $\frac{2}{3}T$. Again, according to some interpretations, present observations tend to indicate that nebulae are constantly being formed and that very many young nebulae are in existence. Should this opinion (which is at present not very widely held) be proved correct, then Lemaître's model will have to be abandoned, and none of the relativistic models is likely to fit such a situation.

Some authors (Eddington, 1939; Omer, 1949; Tolman, 1949) have suggested that general relativity without the cosmological principle should be used in cosmology. Such an approach seems to be quite unjustifiable on present evidence, since it raises more questions than it solves. Without the cosmological principle new explanations would have to be found for the red shift-magnitude relation and for the observed high degree of homogeneity of the universe.

10.9. The previous two sections have discussed the situation in relativistic cosmology as it was in 1950. Many of the considerations then of importance have been superseded by modern developments. In particular the Baade-Sandage revision of the time scale (p. 39) has eliminated the direct conflict between T and other age determinations. Modern work on the origin of the elements (p. 57) has removed some of the attraction of point-source models. Lemaître (1958) has considered his model in the light of these developments, and Klein (1958) has studied various aspects of the point-source question, but there is now room for the examination of other relativistic models.

Oscillating models (Class V, p. 86) have been considered by Wheeler *et al.* (1956), but the serious difficulties connected with Olbers' paradox and the nature of thermodynamics in the contracting phase do not seem to have been examined, let alone resolved.

The comparison between theory and observation is also likely to follow different paths in future (see Chapter XIV) especially since optical number counts are now considered unlikely to be of much help.

CHAPTER XI

KINEMATIC RELATIVITY

11.1. The theory to be discussed in this chapter differs radically in outlook and method from those described in the preceding two chapters. Whereas there an attempt was made to describe the universe in terms of known physical theories, in kinematic relativity the procedure is reversed. The aim of this discipline is to deduce as much as possible merely from the cosmological principle and the basic properties of space, time and the propagation of light. The beauty of this, as indeed of any deductive theory, rests on the rigour of the arguments and the small number of the axioms required.

The theory with its recent developments now covers not only cosmology but a great part of theoretical physics, and the extent of its achievements is greatly to be admired. No less so is the searching analysis to which it has subjected the concepts of time and of space. It has greatly increased our understanding of these subjects.

The findings of the theory are sometimes strange when viewed in the light of ordinary physics, but the great difference in the starting-point suggests that the agreements rather than the disagreements are to be considered remarkable.

When the theory was first developed it met with great hostility and was criticized very severely, often unjustly, and sometimes frivolously. This attitude reacted on the authors of the theory. They learnt by bitter experience that it was useless to spend time in defending their work against these attacks. Unfortunately, this meant that they tended to disregard all criticisms whether justified or not. Such an atmosphere was particularly unhappy for a deductive theory. When an attempt is made to deduce from first principles a fact already known from ordinary physics, only the frankest discussion between proponents and critics can lead to a formulation in which it is clear that the axiomatic deduction is strict and no empirical argument has been slipped in subconsciously.

This historical development has left the theory in a curiously unfinished state. In various places it is plain that empirical facts have been used in the arguments as presented, but one feels that there may, at least in some places, be no need for this appeal to observations. Further investigation and reformulation might well lead to a much stricter line of argument, but, presumably owing to the climate of opinion described, these questions have not been pursued. Accordingly, there are wide divergences of opinion as to which parts of the theory are purely deductive. Everyone is agreed that the cosmological principle, together with the basic facts of the dimensionality of space and time and the fundamental nature of the propagation of light, constitute the basic axioms. But where in the further development a theorem follows clearly from these axioms, where the development though still strictly axiomatic has been diverted from its 'natural' direction by the originator's knowledge of physics, and where a development is not (and possibly cannot be made) rigorously deductive but must be considered as an axiom, all these are questions that have been barely investigated and on which opinions differ widely. The views expressed here are necessarily those of the present author and seem to be intermediate between those of the adherents of kinematic relativity and those of its most critical opponents.

11.2. As has been said before, the theory is based on the cosmological principle in its narrow sense. Hence a very strict and primitive formulation of the principle is required. As Milne (1935, 1948) pointed out, the distinction between kinematic and general relativity becomes very clear at this stage already. In general relativity the equivalence of the *laws of nature* is postulated for *all* observers whatever their position and their state of motion. Such a formulation is clearly inappropriate in kinematic relativity. At the stage at which the cosmological principle is required laws of nature have not been defined. As was discussed in Chapter 1, the distinction between the actual motions and the laws of motion can hardly be drawn in cosmology and is presumably quite inappropriate. The cosmological principle has therefore to be defined with respect to the actual aspect of the universe rather than with respect to the laws of nature. Now a good deal is known from purely terrestrial physics about observers at the same place in different states of motion. From this we know that although the laws of nature may be the same, the aspect of the universe will be different for observers in the same place in different states of motion. Accordingly, it is impossible to define the cosmological principle as implying that all observers, irrespective of their position and state of motion, obtain the same aspect of the universe. It becomes necessary to select a set of 'fundamental observers' of whom there is always only one at each point. The cosmological principle, the contents of which will be discussed later, applies only to this set of observers.

These fundamental observers here introduced are in many ways equivalent to Weyl's bundle of geodesics with respect to which the cosmological principle is introduced into relativistic cosmology. Milne furthermore immediately identified the fundamental particles with the nuclei of the galaxies, but this step seems to deserve a good deal of further examination. In the first instance nothing has been said at this stage about whether the set of fundamental observers is discrete or continuous, and, according to Milne, they form a continuous set. If this is so only some of the fundamental particles can be identified with the nuclei of the nebulae. This raises considerable problems of interpretation in the later development of the theory, problems which so far have hardly been mentioned and certainly not fully discussed.

Another question that appears already at this stage is that of the actual inhomogeneity of the universe and in particular of the random motions of the nebulae. What is the place of these deviations from the detailed application of the cosmological principle in the system of fundamental particles? This question is of very great significance to any theory that relies on the cosmological principle so fully. In his last book Milne expressed the view that the system of fundamental particles (the substratum) plays the part of an imaginary homogeneous background, a sort of universal system of reference against which the inhomogeneities and the random motions have to be considered. This opinion is possibly not applied completely consequentially in the interpretation of the later developments of the theory. Different attitudes to the question are certainly possible, and are felt to be more satisfactory in some authors' views.

11.3. In order to give meaning to the cosmological principle, further definitions and axioms are necessary. Here the searching analysis of the time concept is carried out, which is possibly the greatest achievement of the theory.

The chief concept is that of the passage of time. It is assumed (surely a very modest assumption) that each observer is aware of the passage of time in the sense that he can say of any two *local* events which of them was earlier and which was later. In other words, the events that take place at any one observer form an ordered sequence. A further minor assumption states that the real numbers, themselves an ordered sequence, can be used to attach labels to these events. An earlier event is labelled by a lower number than a later event, but, apart from that, the labelling process is quite arbitrary. It is evidently possible to pass from one system of labelling to another equally valid one by substituting a monotonically increasing function of the old labels. Any system of labelling is called a clock, and the change from one system to another is called a regraduation of the clock.

The great significance of this concept of time becomes clear when the physical realization of clocks is considered. Any repetitive device, such as an oscillation, an orbit described under gravity, etc., can be used as a clock. But since it is impossible to compare intervals of time occurring at different epochs by laying them side by side we cannot speak of a 'uniform flow of time' or of a 'steady even-running clock' without further consideration. It is therefore by no means obvious that different physical clocks, say an atomic and a dynamical clock, should have the same ratio of their periods at all times. A slow variation in, say, the number of oscillations of the spectral line H_α in one day cannot be ruled out from our experience. A separate time-reckoning belongs therefore to every natural phenomenon. Quantum theory shows that the ratios of the frequencies of the spectral lines are related to the integers and hence (since these cannot change continuously) must be constant. Similarly, one would expect the ratio of two different dynamical clocks to remain constant, but their 'dynamical time' may well vary with respect to the 'atomic time' of spectral lines. It is easily seen that this possibility implies that even the 'pure numbers of nature' ($\gamma m_e m_p/e^2$, e^2/hc, etc.) might not necessarily be constant, but in fact kinematic relativity denies any such variation.

This possibility of the existence of different time-scales is an illuminating achievement of kinematic relativity, and is of the greatest importance not only for that theory but also for the theories discussed in the last two chapters of this book.

11.4. Distance measurements are much less fundamental in kinematic relativity than time measurements. The propagation of light clearly offers a possibility of utilizing observations of time intervals in order to measure spatial distances. The method adopted is the radar method. An observer A sends out a light signal at an instant t_1 (by his clock). This is reflected or replied to by B and arrives back at A at time t_2. Observer A then calls $\frac{1}{2}(t_2-t_1)$, multiplied by the conventional velocity of light, the distance of B from him at epoch $\frac{1}{2}(t_1+t_2)$. This measurement clearly depends on the clock A is using. A regraduation of A's clock which replaces t by the increasing function $f(t)$ necessarily implies a regraduation of all distance measurements.

This method of introducing the distance concept into the theory has been severely criticized by Born as being far removed from all practical or indeed practicable measurements of distances. According to Born (1943) no one has ever received light signals from distant nebulae, nor are such measurements at all possible in the time intervals at our disposal. Although there is no doubt much substance in this criticism, a very effective reply can be made. The classical method of distance measurements taken over by general relativity is the use of the rigid ruler. This nebulous concept, utterly inappropriate in any theory of relativity, is wholly out of place in cosmology. Abstract as Milne's light signal technique may be, at least it uses the same phenomenon (light) as is used in practice, and it also follows the natural rule that information travels only along the light cone. Imperfect as Milne's definition of distance may be, it is very much better than the 'rigid ruler' one used in most other theories. Of course the most desirable procedure would be to define distance as luminosity distance, since that is the way in which it is measured. However, no theory has yet been devised using that concept as a starting-point. Though all theories develop the concept of luminosity distance for the purpose of comparison with observation, none contains it as a primitive element, and this must be counted against them all. Milne's definition of distance, by no means perfect as it is, is probably the best yet devised.

A further spatial type of measurement which is permitted in kinematic relativity is the measurement of angles (especially angles between light rays) by means of the type of instrument exemplified by the theodolite. This seems to be a very reasonable procedure and brings the number of space dimensions into the theory.

The formulation of the cosmological principle adopted by Milne is that the totality of the observations that any fundamental observer can carry out by means of his clock and theodolite is identical with those any other fundamental observer can carry out.

11.5. Milne concentrated first on examining one-dimensional systems of observers so that only clocks but no theodolites are required. The problems arising were first solved by Whitrow and Milne. The first important fundamental problem investigated was whether an observer can so regraduate his clock that in some sense it keeps the same time as the clock of some other specified observer. Having shown that this problem possesses a well-defined solution these authors then examined whether a whole set of observers can agree on a universal time. This problem, too, has a solution, if their relative motions satisfy certain conditions. Then there exists a universal or cosmic time for all these observers. This cosmic time is not unique but admits of regraduation. Since the regraduation affects both time and distance measurements, the variation of the mutual distances of all these observers (termed an equivalence) with time depends on the particular type of clock used. Although there is an infinity of these possible time-scales, two are of outstanding importance, not only in this problem but for the whole of kinematic relativity. One is the scale of so-called t-time in which the relative motion of the observers is non-zero but unaccelerated. Choosing a suitable zero of time, the relative distance of any two observers is simply proportional to this t-time. This is easily shown to be equivalent to the velocity-distance relation $v = r/t$. The second time-scale is the so-called τ-time in which the observers appear to be at rest. The transformation from t-time to τ-time is

$$\tau = t_0 \log(t/t_0) + t_0, \tag{11.1}$$

where t_0 is a constant the significance of which is that when $t = t_0$, t-time and τ-time agree both in value and in rate. The zero of t-time is a fundamental event, since it is the instant at which the separation of the observers vanishes. It is the origin of the whole system. In τ-time this event takes place in the infinite past.

Milne preferred to use t-time as his fundamental scale, since it does not involve the constant t_0, but the theory can equally well be built up on the basis of treating τ-time as more fundamental.

In Milne's approach the next step was to deduce that the transformation from one fundamental observer to another is the Lorentz transformation. This beautiful result becomes clearer when it is remembered that the mutual accelerations vanish in the t-system. Since the Lorentz transformations deal chiefly with electromagnetic phenomena, Milne suggests that these follow essentially the t-scale of time. According to this view, an electromagnetic clock would indicate t-time, and since this presumably applies to atomic clocks (spectral lines), t-time is also referred to as atomic time. It will later become apparent that τ-time is more closely associated with dynamical processes.

Milne attempted to carry over the whole argument to three-dimensional equivalences, but it seems that further assumptions are required at this stage in order to specify the curvature of the three-dimensional space. The choice made was that, in t-time, the transformation between any two fundamental observers should still be the Lorentz transformation. This implies that the four-dimensional space is flat, and it can be shown then that not only is the motion of the observers relatively unaccelerated but their three-dimensional space must be hyperbolic.

By using a very different approach Robertson (1935, 1936) and Walker (1936) have shown that it follows from the fundamental assumptions of the theory that it is possible to choose coordinates $(\bar{t}, \bar{x}, \bar{y}, \bar{z})$ and introduce a metric

$$ds^2 = d\bar{t}^2 - R^2(\bar{t})\frac{d\bar{x}^2 + d\bar{y}^2 + d\bar{z}^2}{[1 + \frac{1}{4}k(\bar{x}^2 + \bar{y}^2 + \bar{z}^2)]^2} \tag{11.2}$$

identical in form with the line element of relativistic cosmology (10.8) but rather different in interpretation. For since $\bar{t}$ is still arbitrary, the only significance of (11.2) is that the fundamental observers have fixed $(\bar{x}, \bar{y}, \bar{z})$ coordinates and that the null geodesics of (11.2) represent the light rays. Robertson and Walker both employed methods of the theory of groups to show that (11.2) was the most general line element satisfying the conditions of isotropy and homogeneity required by the cosmological principle.

The time coordinate in (11.2) can be regraduated, since any regraduation leaves the null geodesics unaltered and does not affect the co-moving property of the space coordinates. In order to identify the time coordinate of (11.2) with Milne's t-time it is necessary to regraduate it in such a manner that the motion becomes uniform, that is to say, so that $R(t) = t$. This is always possible, but the resulting four-dimensional space is flat only if $k = -1$, that is, if the 3-space is hyperbolic. The great advantage of this special choice is the applicability of the Lorentz transformation, which is vastly simpler than the corresponding transformation in any other case. It is important to realize, however, that it is this assumption which, in the first instance, separates kinematic relativity from the steady-state theory to be discussed in the next chapter.

The metric (11.2) therefore takes the form

$$\begin{aligned} ds^2 = d\bar{t}^2 - \bar{t}^2 \frac{d\bar{x}^2 + d\bar{y}^2 + d\bar{z}^2}{[1 - \frac{1}{4}(\bar{x}^2 + \bar{y}^2 + \bar{z}^2)]^2} &= dt^2 - dx^2 - dy^2 - dz^2 \\ &= e^{2\tau/t_0} \left\{ d\tau^2 - t_0^2 \frac{d\bar{x}^2 + d\bar{y}^2 + d\bar{z}^2}{[1 - \frac{1}{4}(\bar{x}^2 + \bar{y}^2 + \bar{z}^2)]^2} \right\}, \end{aligned} \tag{11.3}$$

where the first form uses co-moving coordinates and a universal time, the second locally defined measures of time and space in t-time, and the third locally defined measures of space and time in τ-time. The transformation laws are

$$t = \bar{t}\frac{1 + \frac{1}{4}(\bar{x}^2 + \bar{y}^2 + \bar{z}^2)}{1 - \frac{1}{4}(\bar{x}^2 + \bar{y}^2 + \bar{z}^2)}, \quad x = \frac{\bar{x}\bar{t}}{1 - \frac{1}{4}(\bar{x}^2 + \bar{y}^2 + \bar{z}^2)}, \quad \tau = t_0 \log(\bar{t}/t_0). \tag{11.4}$$

Since the coordinates $\bar{x}$, $\bar{y}$, $\bar{z}$ are constant for any observer, it follows that x/t is constant. Hence the velocity-distance relation takes the simple form

$$u = \frac{dx}{dt} = \frac{x}{t}, \tag{11.5}$$

or, in vector form,

$$\mathbf{v} = \frac{\mathbf{r}}{t}. \tag{11.5'}$$

The completion of Milne's model of the substratum required a determination of the variation of density with time. It follows readily from the theory of the Lorentz transformation that not only ds but also the quantity

$$X = t^2 - x^2 - y^2 - z^2 \tag{11.6}$$

is invariant, i.e. independent of the observer. Accordingly, the invariant density ρ, that is, the number of particles in the invariant volume element $dx\,dy\,dz\,dt/ds$, can only be a function of X. The apparent density n is the number of particles in $dx\,dy\,dz$. Accordingly, we have

$$n = \rho(X)\,dt/ds. \tag{11.7}$$

Also, by (11.3) and (11.5),

$$\left(\frac{ds}{dt}\right)^2 = 1 - \left(\frac{dx}{dt}\right)^2 - \left(\frac{dy}{dt}\right)^2 - \left(\frac{dz}{dt}\right)^2 = \frac{X}{t^2}. \tag{11.8}$$

Therefore

$$n = t\rho(X)\,X^{-\frac{1}{2}}. \tag{11.9}$$

Milne then made the assumption, which is evidently suggested by ordinary physics, that matter is conserved. This is the second arbitrary though plausible assumption differentiating kinematic relativity from the steady-state theory.

Hence the equation of hydrodynamic continuity applies:

$$\frac{\partial n}{\partial t} + \operatorname{div}(n\mathbf{v}) = 0. \tag{11.10}$$

By (11.5′) and (11.9) this implies

$$\frac{\rho}{X^{\frac{1}{2}}} + 2t^2 \frac{d}{dX}\left(\frac{\rho}{X^{\frac{1}{2}}}\right) + 3\frac{\rho}{X^{\frac{1}{2}}} - 2\mathbf{r}^2 \frac{d}{dX}\left(\frac{\rho}{X^{\frac{1}{2}}}\right) = 0. \tag{11.11}$$

Hence we find that

$$\rho = BX^{-\frac{3}{2}}, \quad n = BtX^{-2}, \tag{11.12}$$

where B is a constant.

The meaning of (11.5′) and (11.12) is quite clear: The velocity-distance relation applies strictly, the density decreases in time but increases from the origin towards $X = 0$. The surface $X = 0$ represents the invariant border of the universe which is advancing at the speed of light.

This completes the discussion of the substratum in kinematic relativity. Since no random motions (heat) have been introduced the question of pressure does not arise. Although the further developments of the theory are of the greatest importance and interest, it may be worth pointing out that most of the other theories do not as such go beyond the construction of a substratum. In the theories discussed in the preceding chapters the local effects are treated by the local forms of the theories concerned without making use of the cosmological principle. In other theories (discussed in subsequent chapters) little or no progress has as yet been made in contemplating any further details of structure. But kinematic relativity, aided no doubt by the particularly simple and lucid structure of the substratum following from its assumptions, continues to base itself on the cosmological principle in order to construct a dynamics and other branches of theoretical physics.

11.6. The first step, and in many ways the most important and characteristic one, in the further development of the theory concerns the equation of motion of a 'free particle'. The term 'free' denotes the opposite of fundamental; that is, we are now concerned with particles not partaking of the motion of the substratum. Of course a further assumption has to be made before the problem can be examined; for there is no *a priori* reason why there should be any equation describing the motion of a free particle. The assumption made is the very plausible one of accepting Galileo's principle that the acceleration is a unique function of position, velocity and epoch. This last independent variable must clearly be included in any changing universe.

The problem is therefore as follows: A given fundamental observer discovers that for every free particle the acceleration $\mathbf{f} = d^2\mathbf{r}/dt^2$ is a definite function of $\mathbf{v} = d\mathbf{r}/dt$, $\mathbf{r}$ and t, so that

$$\mathbf{f} = \mathbf{g}(\mathbf{v}, \mathbf{r}, t). \tag{11.13}$$

Applying now the cosmological principle it appears that although for any other fundamental observer $\mathbf{f}$, $\mathbf{v}$ and $\mathbf{r}$ have to be transformed, the functional form of $\mathbf{g}$ must be independent of the observer.

The solution of the problem presents no great difficulties, since the Lorentz transformations of acceleration are easily derived, but the algebra becomes rather heavy. The result is

$$\mathbf{f} = (\mathbf{r} - \mathbf{v}t)\frac{Y}{X}G(X, \xi), \tag{11.14}$$

where

$$X = t^2 - \mathbf{r}^2, \quad Y = 1 - \mathbf{v}^2, \quad Z = t - \mathbf{v}.\mathbf{r}, \quad \xi = \frac{Z^2}{XY}, \tag{11.15}$$

while G is an arbitrary non-dimensional function of its arguments.

At this stage Milne used a remarkable argument characteristic of the whole theory. Originally he called it the 'dimensional hypothesis', but later did not consider that any hypothesis was involved at all. Briefly, the argument is that since (11.14) involves a property of such a simple system as the substratum which was built up without reference to any fundamental dimensional constants, no such constant can appear at this stage. Accordingly, the non-dimensional function G cannot depend on a dimensional argument like X. Therefore $G = G(\xi)$. The great significance of this step will become apparent later.*

The vectorial part of (11.14) determines the direction of the acceleration **f**, and it is readily seen that this is along the line joining the free particle to that fundamental particle having the same velocity as the free particle. Since X and Y are positive, the acceleration is towards that fundamental particle if G is negative and away from it if G is positive. In the first case the system is essentially stable in that there is a tendency for the velocities of free particles relatively to the local fundamental particles to diminish, in the second case it is correspondingly unstable. In fact Milne was able to show that $G \leqslant -1$.

11.7. A further development is now possible. Suppose that a whole swarm of free particles of different positions and velocities is superposed on the substratum. Can the structure of this assembly be such that it, too, satisfies the cosmological principle in that it presents the same aspect to every fundamental observer?

* While this step is justified in the construction of an ideal universe it by no means follows that formulae so derived apply in the actual universe in which dimensional constants are defined by the elementary particles.

To put this more precisely, consider the distribution of particles in phase space. This involves the setting up of a distribution function $f(x, y, z; \dot{x}, \dot{y}, \dot{z}; t)$ such that the number of particles in the intervals $x \pm \frac{1}{2}dx, \ldots; \dot{x} \pm \frac{1}{2}d\dot{x}, \ldots$ at time t is $f\,dx\,dy\,dz\,d\dot{x}\,d\dot{y}\,d\dot{z}$. A distribution function like this is significant only if it is assumed, as has been done above, that the acceleration is determined by position, velocity and epoch.

It is easily deduced from the Lorentz transformations that $Y^{-\frac{5}{2}}$ times the volume element of phase space $dx\,dy\,dz\,d\dot{x}\,d\dot{y}\,d\dot{z}$ is an invariant. Since the only other invariants are X and ξ it follows that

$$f = Y^{-\frac{5}{2}}\Phi(X, \xi), \qquad (11.16)$$

where Φ is an arbitrary function of its arguments of dimension $(\text{length})^{-3}$. Hence $X^{\frac{3}{2}}\Phi$ is non-dimensional, and Milne's dimensional argument can again be employed. Then

$$f = Y^{-\frac{5}{2}}X^{-\frac{3}{2}}\Psi(\xi). \qquad (11.17)$$

The law of conservation of mass (particle number) may now be applied in phase space, the acceleration being given by (11.14). The result is a connexion between the acceleration function $G(\xi)$ and the distribution function $\Psi(\xi)$ which takes the form

$$G(\xi) = -1 - \frac{C}{(\xi - 1)^{\frac{3}{2}}\Psi(\xi)}, \qquad (11.18)$$

where C is a constant of integration.

Milne interpreted (11.18) as a law of gravitation, since it determines the accelerations in terms of the distribution of the masses (particles). Of course it is not a general law of gravitation, but applies only to mass distributions satisfying the cosmological principle.

Milne proceeded to identify the first term in (11.18) with the gravitational effect of the substratum alone, and the second term, which depends solely on Ψ, with the gravitational effect of the statistical distribution of free particles. Although the result is plausible none of the several proofs given by Milne can be considered wholly convincing. At the present time this identification has therefore the status of a working hypothesis, but it may well be possible to prove it rigorously in the future.

In the absence of free particles, (11.14) therefore takes the form

$$\frac{d\mathbf{v}}{dt} = -\frac{Y}{X}(\mathbf{r}-\mathbf{v}t). \qquad (11.19)$$

It is of interest to transform this formula to τ-time. Since $d\tau/dt = t_0/t$, and since distances are measured by the lapse of time, it follows that for near objects ($\mathbf{r}/t$ small) the distance Π in τ-time is

$$\Pi = \frac{t_0}{t}\,\mathbf{r}. \qquad (11.20)$$

Hence
$$\mathbf{v} = \frac{d\mathbf{r}}{dt} = \frac{1}{t_0}\,\Pi + \frac{d\Pi}{d\tau} = \frac{\mathbf{r}}{t} + \frac{d\Pi}{d\tau}. \qquad (11.21)$$

Furthermore, for small $\mathbf{v}$ we have $Y \doteqdot 1$, $X \doteqdot t^2$, and therefore (11.19) takes the form

$$\frac{d\mathbf{v}}{dt} = -\frac{\mathbf{r}-\mathbf{v}t}{t^2},$$

or
$$\frac{d}{dt}\left(\mathbf{v}-\frac{\mathbf{r}}{t}\right) = 0. \qquad (11.22)$$

By (11.21) this implies
$$\frac{d^2\Pi}{d\tau^2} = 0. \qquad (11.23)$$

Hence, for near, slowly moving objects, in the absence of other free particles, Newton's first law of motion applies in τ-time, which Milne hence identified with the time of Newtonian dynamics. Although in a comparison with Newtonian dynamics the restriction to near, slowly moving particles is fully justified, it is of interest to note that a strict treatment for general particles leads to an expression which is the closest parallel to (11.23) in the hyperbolic space $\tau = \text{const.}$ (cf. (11.3)).

Milne and Walker carried out a very full examination of (11.18). In their view (11.18) leads not only to Newton's first law, but, with appropriate interpretations, leads to an inverse square law of gravitation. The constant C turns out to give the relation between particle *numbers* (only these enter Ψ) and their effective gravitational masses.

This remarkable derivation of the fundamental laws of Newtonian dynamics from the structure of the substratum undoubtedly constitutes a great achievement for kinematic relativity and indicates the great power of its methods. Encouraged by these successes Milne and Walker built up a complete system of dynamics. Although most of its results are (as they must be) in agreement with Newtonian theory there are one or two deviations which, though small on the local scale, are of great importance for cosmic problems. Probably the most significant one of these is a secular increase in angular momentum, a quantity which turns out to be proportional to the cosmic time t. This result greatly aided Milne's theory of nebular structure, which also leads to a dynamical theory of their spiral structure.

Further developments of kinematic relativity lead, beyond dynamics, to new theories of the photon and electromagnetic fields, and even supply a new basis for atomic and nuclear theory. The formulation of the photon theory seems to be somewhat in doubt at present. There is no space here for a description of these fascinating theories which, though at present certainly guided by (and probably in one or two places even based on) a knowledge of ordinary physics, may possibly one day be formulated in a purely deductive form.

11.8. The foregoing brief description will have indicated the remarkable success of kinematic relativity in attempting to use the cosmological principle not only for the construction of the substratum but as chief guide in formulating ordinary physics. In this respect it differs greatly from all other cosmologies which either rely on a conventionally obtained body of physics or have not yet succeeded in drawing conclusions of local interest from the cosmological principle. But it must be pointed out that, however fascinating and successful kinematic relativity may be as an abstract discipline, grave difficulties stand in the way of interpreting it as a valid description of our universe.

One type of difficulty is directly associated with the fundamental axiom of the theory, the cosmological principle. Without a strict (as opposed to statistical) formulation of this principle, the mathematical complexities of constructing the theory would be enormous, but this very strict formulation leads to awkward problems of interpretation, which have not so far received much attention.

In the first place Milne considered the substratum as an ideal concept serving as a universal system of reference, supplying everywhere a local standard of rest against which any local deviations must be considered. In agreement with this interpretation Milne proved that the fundamental particles form a continuous system.

However, in another place, he identified the fundamental particles with the nebular nuclei. Although this interpretation gives a great deal of substance to his picture of the substratum (which would otherwise be a purely abstract concept) it raises difficulties. The nebular nuclei form in fact a discrete system. Does this mean that only some of the fundamental particles have 'materialized' into nebular nuclei? If it were so, a conflict would arise not only with Milne's earlier interpretation but also with his strict formulation of the cosmological principle. Similar problems are raised by the observed individual motions of the nebular nuclei.

Possibly even more severe is yet another difficulty of this type. Milne deduced from (11.18) that in order to avoid singularities of $G(\xi)$ it is necessary for $\Psi(\xi)$ to become very large near $\xi = 1$. This implied a strong tendency for the free particles to exist closely grouped round the fundamental particles, and these agglomerations of free particles were, according to Milne, the nebulae themselves. But what can these agglomerations mean if the fundamental particles form a continuous set? Concentrations of free particles surrounding *every* point in space appears to be a meaningless concept and can hardly be identified with the nebulae themselves.

A possible line of argument supporting Milne's view would be that the formulation of the cosmological principle was made overstrict for mathematical convenience, and that a 'loosening' of the basis would lead to a 'loosening' of the results sufficient to clarify the comparison with observation. Such a supposition is evidently very dangerous, but, on the other hand, the difficulties standing in the way of basing a theory on a statistical cosmological principle are very great indeed. If this supposition were stated clearly and some reasons given making it plausible, and if, furthermore, it were explained that owing to this difficulty comparisons between theory and observation were bound to be provisional and hypothetical, then this view would command sympathy. However, the present tendency to gloss over this point is bound to raise doubts.

The photon theory of kinematic relativity leads to a very interesting result. Its expression for the luminosity distance differs from the relativistic one in that the Doppler-shift factor appears only singly and not squared. Hubble and Tolman, in their analysis of the nebular counts, found that it gave results which they found much more satisfactory if only a single Doppler-shift factor was allowed for. Although their analysis has been criticized, and although the uncertainties of the data are substantial, this result may be considered to be favourable to kinematic relativity.

In τ-time the nebulae do not move, i.e. the universe presents a static appearance. The red shift of the spectral lines of distant objects is not due to a velocity of recession but to the fact that the frequency of light (though constant in t time) decreases in the course of time in the τ-picture. This temporal decrease of frequency is one way of overcoming Olbers' paradox and clearly leads to only one red shift factor in the luminosity formula.

11.9. The relation between the old value of Hubble's constant and Milne's model of the universe, known as the time-scale problem, is of great historical interest. The observations have been discussed in Chapters V and VI. It is evident that the recession velocities of the nebulae lead to a determination of the value of Milne's cosmic or atomic time t at the present epoch, and that this value of t was believed to be about $1{\cdot}8 \times 10^9$ years.

In the mid-thirties when kinematic relativity was first developed the only other determinations of the age of the universe were Jeans' arguments (1928) based on the *dynamics* of proper motions of stars and of the orbits of binaries.* These arguments, which have since been shown to be wholly fallacious, led to an age of the galaxy of about 10^{12} years. The time-scale difficulty was therefore extremely severe at that time. Milne's theory, however, resolved it completely, since its dynamics was controlled by the dynamical time τ which owing to its logarithmic dependence on t went back to the infinite past. In other words, Milne's dynamics showed that in the distant past dynamical processes were much faster (when

* The estimates of the age of the Earth were then already close to the present estimates (cf. Jeffreys, *The Earth*, 2nd ed., pp. 64–73). They were, however, considered to be of no relevance to cosmology since they were much shorter than Jeans' estimates.

viewed in t-time) than they are now, and hence there was no conflict between Jeans' and Hubble's data.

Present-day determinations of the time-scale depend largely on nuclear processes. Since they lead to ages of the order of 5×10^9 years, and modern determinations of Hubble's constant lead to a present value of t around $1{\cdot}3 \times 10^{10}$ years, the conflict has disappeared.

CHAPTER XII

THE STEADY-STATE THEORY

12.1. The steady-state theory (Bondi and Gold, 1948) differs from the theories so far discussed in that the problem of the origin of the universe, that is, the problem of creation, is brought within the scope of physical inquiry and is examined in detail instead of, as in other theories, being handed over to metaphysics. This achievement is accomplished at a price, a modification of the law of conservation of mass, but this apparently drastic step is shown not to lead to any conflict with observation or experiment. Originally the theory was also believed to overcome the time-scale difficulty which was then so great for other theories, but owing to the revision of the observational results this consideration is now irrelevant.

The fundamental assumption of the theory is that the universe presents on the large scale an unchanging aspect. Since the universe must (on thermodynamic grounds) be expanding, new matter must be continually created in order to keep the density constant. As ageing nebulae drift apart, due to the general motion of expansion, new nebulae are formed in the intergalactic spaces by condensation of newly created matter. Nebulae of all ages hence exist with a certain frequency distribution. Astrophysical estimates of the age of our galaxy do not put it into a very rare class of nebulae.

The theory is deductive in the sense that its conclusions are derived from the cosmological principle, but the very powerful formulation of the principle employed dispenses with the need for additional assumptions.

A different approach has been proposed by Hoyle and will be discussed later in this chapter. In that formulation a suitable modification of the field equations of general relativity is taken as starting-point, so that the conclusions reached are very similar to those of the steady-state theory.

The importance to the theory of the powerful formulation of the cosmological principle makes it highly desirable to examine in detail the arguments for the acceptance of this principle. This examination reveals that the arguments supporting the usual narrow cosmological principle imply the validity of the wider perfect cosmological principle, according to which the large-scale aspect of the universe should not only be independent of the *position* of the observer but also of the *time* of making the observation.

The arguments leading to this formulation have already been discussed briefly in Chapter II, but owing to their importance for the theory they will be recapitulated. The entire subject of cosmology rests on the assumption that our terrestrially obtained knowledge of the physical world can be applied to the universe at large. This assumption is but another expression of the Copernican point of view that there is nothing exceptional about the Earth, but that it and its neighbourhood are 'typical'.

Now this assumption of ability to apply our knowledge necessitates a similarity of circumstances. It would be quite an unwarranted extrapolation of the physical sciences to say that they would apply even if we were embedded in a heat bath of 10^8 degrees, or if we lived in a world where inertia was less but gravitation and electricity more powerful than in our world. But as soon as it is assumed that the universe looked different at some place or some time from its present aspect here, we are forced to attempt to apply physics in such strange and novel circumstances, of the effects of which we are wholly ignorant. The strong connexions which apparently exist between the universe at large and local physics were emphasized and illustrated in Chapters III, IV and V. The effects of any variation in the universe are wholly unknown. There is, however, a possibility of avoiding the need to consider this difficult subject. If the universe presents the same aspect to every fundamental observer, wherever he is and at all times, then none of these difficulties and doubts arises. The assumption that this is the case is the 'perfect cosmological principle' fundamental to the steady-state theory. If it is correct, then there are no ambiguities in applying physics anywhere at all times. If it should turn out to be false then progress in cosmology will be infinitely more difficult to achieve. The perfect cosmological principle is a clear straightforward principle of great power and leads to a particularly simple picture of the universe. Great stress is laid on the fact that no ambiguities or difficulties of interpretation arise as is so frequent in other theories. Problems such as, for example, the connexion between nebular evolution and the number counts do not exist, since, by virtue of the perfect cosmological principle, the age distribution of the nebulae is always the same.

The merits of this theory, in the first instance, are its simplicity and directness, but it will be seen that it also solves the time-scale difficulty, the grave of so many theories, and deals with various other problems. Some of the conclusions of the theory appear to differ from current concepts, though they nowhere disagree with observation.

In any theory in which there is a changing universe, the variation of physics consequent upon this change has to be fixed arbitrarily, being wholly outside the domain of experience. General relativity makes the arbitrary assumption that its field equations apply in spite of changes in the universe; kinematic relativity leads to spectacular changes in the physical constants, but again these are the result of arbitrary assumptions.

Although the perfect cosmological principle states that the universe presents an unchanging aspect and is hence in a steady state, this in no way implies that the universe has to be static and motionless. The distinction between a steady and a static state is well known from hydrodynamics. A river may be in a steady state in that the velocity of the water is a function of position only and not of time. It will then present a stationary aspect to any observer at rest, but not only may the water be moving, but each particle will in general even suffer accelerations or retardations as it moves from regions of low velocity to regions of high velocity or conversely.

12.2. The most immediate consequences of the perfect cosmological principle have already been discussed in Chapter III. They concern the present state of thermodynamic disequilibrium of the universe. Since in a static unchanging universe thermodynamic equilibrium will eventually be set up, the perfect cosmological principle in conjunction with experience shows that our universe cannot be static. In a contracting universe the Doppler shift leads to a disequilibrium in which radiation preponderates over matter, whereas the opposite is true in an expanding universe. Accordingly, the steady-state theory, alone amongst all theories, deduces the fact that the universe is expanding from the local observations of thermodynamic disequilibrium. The observations of the recession of distant nebulae then merely serve as a check on the theory. Since the linking of different observational facts is one of the chief purposes of any scientific hypothesis, the discovery of the connexion just discussed is a particular achievement of the steady-state theory.

12.3. The next deduction to be made from the perfect cosmological principle has formed the most controversial point of the theory. The expansion of the universe, which can be inferred either from thermodynamics or from astronomical observations, would seem to lead to a thinning out of material. By the perfect cosmological principle the average density of matter must not undergo a secular change. There is only one way in which a constant density can be compatible with a motion of expansion, and that is by the *continual creation of matter*. Only if the diminution of density due to the drift to infinity is counteracted by a constant replenishment of newly created matter can an expanding universe preserve an unchanging aspect.

There is little doubt that the continual creation of matter necessary in this theory is the most revolutionary change proposed by it. There is, however, no observational evidence whatever contradicting continual creation at the rate demanded by the perfect cosmological principle. It is easily seen that this is, on the average,

$$3 \times (\text{mean density of matter in the universe}) \times \text{Hubble's constant} = 10^{-46}\ \text{g./cm.}^3\text{sec.}$$

approximately. In other words, on an average the mass of a hydrogen atom is created in each litre of volume every 5×10^{11} years. As will be seen later, there are strong arguments showing that the creation rate does not vary widely between different places, so that the average rate given above has universal significance. It is clear that it is utterly impossible to observe directly such a rate of creation. There is therefore no contradiction whatever with the observations, an extreme extrapolation from which forms the principle of conservation of matter. The argument may be stated in the following terms: When observations indicated that matter was at least very nearly conserved it seemed simplest (and therefore most scientific) to assume that the conservation was absolute. But when a wider field is surveyed then it is seen that this apparently simple assumption leads to the great complications discussed in connexion with the formulation of the perfect cosmological principle. The principle resulting in greatest overall simplicity is then seen to be not the principle of conservation of matter but the perfect cosmological principle with its consequence of continual creation. From this point of view continual creation is the simplest and hence the most scientific extrapolation from the observations.

In the steady-state model the amount of matter is constant in the part of the universe observable with a telescope of given power, that is, in the part within any fixed distance from the observer. In this sense matter is conserved in any constant *proper volume* of space. In the relativistic models, on the other hand, the amount of matter in a similarly defined part of the universe is diminishing, since the density of matter per unit *coordinate volume* (in the usual co-moving coordinates) is constant. It may well be considered more correct to speak of conservation of mass in the steady-state model rather than in relativity, since proper volume is more fundamental than coordinate volume.

It should be clearly understood that the creation here discussed is the formation of matter not out of radiation but out of nothing. The expansion of the universe leads, through the operation of the Doppler shift, to a continual loss of radiation. The average density of radiation must stay constant by virtue of the perfect cosmological principle and is replenished by the action of matter in stars which constantly generates fresh radiation. In this way the creation process, together with the expansion of the universe, prevents the approach of the heat death, the state of thermodynamic equilibrium in which no evolution can take place and in which the passage of time has no significance. High-entropy energy (in the form of radiation) is constantly being lost through the operation of the Doppler shift in the expanding universe, while low-entropy energy is being supplied in the form of matter.

12.4. Before proceeding to discuss the physics of the creation process in more detail it is of interest to discuss the relation of the perfect cosmological principle to astronomical observations. What is the overall picture of the stationary expanding universe? Which of the many models previously discussed are compatible with the perfect cosmological principle? To what extent are they defined by it, without the use of theories incompatible with the creation process, such as Newtonian dynamics or general relativity?

Fortunately, the achievements of kinematic relativity point the way. For the smoothed-out model of the universe (the substratum) far-reaching deductions can be made from the kinematical consequences of the cosmological principle. As was said in Chapter XI, this type of problem has been thoroughly investigated by Robertson and by Walker. They rely only on:

(i) the cosmological principle in its narrow form,

(ii) the fundamental properties of the propagation of light,

(iii) the definition of the substratum including the assumption (Weyl's postulate) that the particle paths do not intersect except possibly at one singular point in the past.

They find that under these assumptions the universe can be plotted on a Riemannian map with metric

$$ds^2 = dt^2 - R^2(t)\,[dr^2 + r^2\,d\theta^2 + r^2\,\sin^2\theta\,d\phi^2]/[1 + \tfrac{1}{4}kr^2]^2, \quad (12.1)$$

in which the fundamental particles have constant (r, θ, ϕ) coordinates, t measures their proper time, and light rays are null-geodesics. The quantity $R(t)$ is an arbitrary function of the time and k is a positive, zero or negative constant. This is just the Robertson line element (cf. (10.8)) of the models of relativistic cosmology.

The stationary property of the model of the steady-state theory has not yet been used. Since the universe is expanding, $R(t)$ must be an increasing function of the time. Now the square of the radius of curvature of the (r, θ, ϕ) 3-space is kR^{-2}. It is an observable quantity, since it affects the rate of increase of the number of nebulae with distance, and must hence be constant by virtue of the perfect cosmological principle. But R varies with time and hence we must have $k = 0$.

The function $R(t)$ itself is not directly observable, since its value depends on the units of r. It is true that the variation of $R(t)$ implies that the distance of every individual nebula varies with time, but this agrees with observation and certainly does not conflict with the perfect cosmological principle, which only states that the *aspect* of the universe is steady, not that individual nebulae remain fixed. Now $\dot{R}(t)/R(t)$ is the constant of the velocity-distance law (Hubble's constant, cf. (9.13)), and this observable quantity, characteristic of the whole motion of the nebulae, must remain constant. Accordingly $R(t)$ is an exponential and hence the metric is

$$ds^2 = dt^2 - [dr^2 + r^2 d\theta^2 + r^2 \sin^2\theta \, d\phi^2] \exp(2t/T), \qquad (12.2)$$

where $1/T$ is Hubble's constant.

This is the metric of de Sitter, which is rejected by general relativity because the field equations imply that it represents empty space. In the steady-state theory this difficulty does not arise because the field equations of general relativity (implying the conservation of matter) are rejected.

It is of interest to note that without any further assumption the perfect cosmological principle leads to the metric (12.2) which is completely specified except for the scale factor T. The de Sitter metric (12.2) is particularly simple and the relations between the observable quantities are easily worked out (t goes from $-\infty$ to $+\infty$, zero of t conventional).

Consider the observations that can be carried out by an observer at $r = 0$, $t = 0$. Along a light track $\theta = \text{const.}$, $\phi = \text{const.}$, $ds = 0$ and hence $dr = \pm dt \exp(-t/T)$. For an incoming ray reaching $r = 0$ at $t = 0$, we have therefore

$$r = T(e^{-t/T} - 1). \qquad (12.3)$$

It may be deduced that the Doppler shift of an object with coordinate r observed at $t = 0$, $r = 0$ is

$$1 + \delta\lambda/\lambda = 1 + r/T, \qquad (12.4)$$

while the intensity of light received is

$$\frac{L}{4\pi r^2(1 + r/T)^2}, \qquad (12.5)$$

where L is the strength of the source.

These formulae are identical with those of relativistic cosmology applied to de Sitter's model, but an important difference arises when the third observable quantity, the number of nebulae in a given range, is considered. As a consequence of the perfect cosmological principle, the number n of nebulae per unit proper volume must be constant; it was the same in the past as it is now, in spite of the expansion of the universe that has taken place since. It can be deduced from (12.2) and (12.3) that the number of nebulae with radial coordinate between r and $r+dr$ from which light reaches $r=0$ at $t=0$ is

$$4\pi r^2 n\, dr(1+r/T)^{-3}. \qquad (12.6)$$

In relativistic cosmology the last factor does not occur. For in relativistic cosmology the number of nebulae per unit *coordinate volume* $r^2 \sin\theta\, dr\, d\theta\, d\phi$ is taken to be constant, whereas in the steady-state theory the number of nebulae per unit *proper volume* is constant. Owing to the expansion of the co-moving system of coordinates, the assumption of relativistic cosmology implies a higher nebular density per unit proper volume in the past than now. The last factor in (12.6), characteristic of the steady-state theory, arises from the relation between coordinate and proper volumes.

A further difference between the theories arises since the perfect cosmological principle implies that both n and the average nebular luminosity L must necessarily be independent of the time, whereas in relativistic cosmology additional assumptions are required to establish the variations of these quantities in time. Hence no such difficulties arise in the steady-state theory as in other theories where a complete theory of nebular evolution is required before the observations of the most distant nebulae can be interpreted, and any discrepancy can always be ascribed to the variations of L and n with time.

Formulae (12.4), (12.5) and (12.6) enable a comparison between theory and observation to be made, the reduction of the data being carried out with due regard to the K-term. The three parameters n, L, T are all fixed by the observations of near nebulae, and the number counts of distant nebulae serve merely as a check on the theory, contrary to what is done in general relativity, where they are used to select a particular model. Nevertheless, the agreement between theory and observation is very good* (Fig. 6). If the lower

* As in Chapter X, 1938 data are used here and so this comparison is of purely historical interest.

value of X (p. 40) is used the agreement is perfect, if the higher one is used the agreement is fairly good. In either case the observations are as well represented by the theory as can be expected in view of the uncertainty of the data. In spite of the great number of their adjustable parameters none of the relativistic models leads to as unforced an agreement as this well-determined model of the steady-state theory. Hubble's original analysis of the data suggested that distant nebulae were so few as to demand an improbably small spherical universe. The theory described in this chapter explains the same fact as due to the unchanging proper density of nebulae.

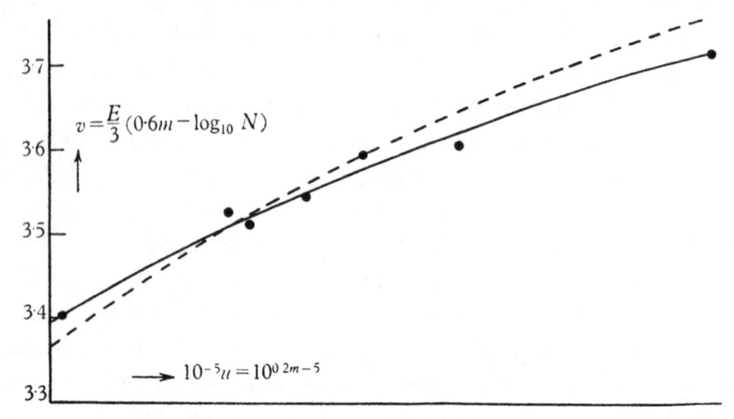

Fig. 6. Comparison between the steady-state theory and observation.
• Nebular counts (Table I, p. 41).
— Theoretical curve for $X = 5$.
--- Theoretical curve for $X = 4{\cdot}7$.
Both curves are drawn for an effective temperature of 5800°.

12.5. Although the average aspect of the universe is unchanging in the steady-state theory, individual nebulae undoubtedly evolve and age. The following picture then arises. Nebulae are condensed out of the intergalactic matter. As time progresses they form stars, the stars age, and so on. All the while nebulae of the same age-class are, owing to the expansion of the universe, drifting apart from each other, but new nebulae are constantly forming in the vast spaces between the older ones where the bulk of the new matter is created. Old nebulae are far from each other and their density of occurrence is low. Presumably, having had time to give effect to their gravitational attraction, they form clusters. In the picture of the universe as a whole, the age of any particular nebula is a purely statistical attribute. The average age (from some definition of the youngest recognizable nebula) is only $\frac{1}{3}T = 4 \times 10^9$ years, but any particular nebula may be arbitrarily old. For example, our galaxy may well be about 10^{10} years old as indicated by various data (Chapter VI). Its relatively large size and the degree of clustering round it indicate quite independently that it may be somewhat older than the average extragalactic nebula. The nearest cluster of similar age should be about twenty times as far from us as the average distance between nebulae, in fair agreement with the observations of the nearest clusters.

There is no point origin and no initial catastrophe in this theory. Accordingly, the origin of the heavy elements must be assumed to be in special stellar processes (supernovae) in agreement with the theory mentioned in Chapter VIII, unless it is assumed that not only hydrogen but all elements are continually created. The second alternative does not seem to be particularly simple or necessary.

12.6. The most novel feature of this theory is the creation of matter. Although this process is not directly observable it is of great interest to discuss the physics of the creation mechanism.

In the first instance it is clear that the creation rate per unit volume per unit time cannot vary very widely from place to place, not nearly as much as, say, the density of matter does. For if the creation rate were, for example, proportional to the density of matter, then the new matter would be created in the interior of the stars at a very high rate (doubling their masses every 3×10^9 years) and not in empty space where it is needed for the formation of new galaxies. For by virtue of the perfect cosmological principle new nebulae must condense in the growing spaces between the older ones. A small degree of variation cannot be ruled out on these grounds. For reasons of simplicity it seems best, at the present stage, to assume that the creation rate is constant in space and time. Hence the amount of matter created in a small four-dimensional element of space-time is proportional to the volume of the element, the factor of proportionality being three times the product of Hubble's constant ($1/T$) and the mean density of matter in the universe (ρ_0).

Therefore the creation of matter must also occur in intergalactic, more or less empty, space. This has an important consequence. For if matter were only created in the presence of other matter then it could be assumed to appear with the velocity of that matter. Creation in empty space implies, however, that a different definition of the initial velocity has to be found which is of universal applicability.

The law that defines this initial velocity cannot be formulated in a Lorentz invariant way since there is no Lorentz invariant velocity, and a Lorentz invariant probability distribution of velocities necessarily diverges. Any finite probability distribution defines a velocity of its mean. Hence the creation of particles necessarily implies a preferred direction in space-time. However much this may contradict orthodox relativity, the concept of a preferred direction is common to all cosmological theories (Weyl's postulate). The steady-state theory differs from other theories only in assigning an immediate physical significance to this direction by identifying the velocity of the substratum (fundamental observers) with the velocity of the newly created particles.

There is no difficulty in defining this preferred direction even in a non-uniform universe. For consider an observer at a certain point in space-time. The aspect the universe presents to him depends on his state of motion, since this affects the Doppler shifts and apparent luminosities of all objects. The difference of the sum of the products of Doppler shift and light energy received from the nebulae in two complementary hemispheres is a monotonic function of the component of the observer's velocity normal to the dividing plane. For a certain velocity the two hemispheres will make equal contributions. Similarly, the other components of his velocity can be adjusted until, in the respect mentioned, the universe appears to be isotropic to the observer. The velocity so defined will be perfectly definite. Isotropic appearance with respect to some other quantity can be defined similarly, but owing to the great uniformity of the universe different properties lead to virtually identical velocities provided only they all assign sufficient weight to distant objects.

There is therefore no difficulty of principle involved in the physical law defining the starting velocity of newly created matter.

The lack of Lorentz invariance is new to a local physical law, but is by no means new to cosmology. On the contrary, it is common to all theories of cosmology whether it is introduced as Weyl's postulate or as the law of motion of fundamental particles. What the steady-state theory does is to link the preferred velocity (which exists everywhere according to all theories) with a definite physical process.

The velocity so defined may be the actual velocity of every newly created particle, or it may only be an average around which the actual velocities are distributed according to some statistical law. The second possibility would correspond to an 'initial temperature'. Which of these two possibilities is correct cannot be said at present. If the initial temperature were high and the random velocities were dissipated in collisions, these might possibly lead to more than the observed radiation from intergalactic space. On the other hand conditions in intergalactic space are little understood. The temperature may be quite low, and there is no argument showing that the simplest picture (zero random velocities) is incorrect.

Similarly, it may be shown from the absence of any observable effects of a net charge, that the average rate of creation of charge must be very low if not nil. The matter is therefore presumably created in the form of neutral hydrogen atoms, but the possibilities of separate creation of electrons and protons or of the creation of neutrons cannot be excluded.

The creation rate, as was stated before, cannot vary too much. It seems simplest to suppose that the probability of creation in any small four-dimensional element of space-time is simply proportional to its four-dimensional volume. The factor of proportionality is either a universal constant or varies slowly in space and time, like, say, total gravitational potential.

The theory offers no explanation of the numerical coincidences, but at least they are permanent, since all the constants of nature remain the same at all times. It is also true that in this theory all the coincidences are purely atomic, since the constant T has not only cosmic significance but also (through its close connexion with the creation constant) direct atomic importance.

At present the equations of motion for inhomogeneities have not been formulated. Therefore the theory must be considered incomplete, but there seems to be an attractive possibility of linking this formulation with Mach's principle. As was stated before, general relativity does not incorporate Mach's principle (at least not in detailed form). The difficulties of expressing a linkage between very distant matter and local physics have been insurmountable as long as it was attempted to connect a *changing* universe with a *permanent* behaviour of matter near us. In the steady-state theory the unchanging character of the universe should greatly simplify such a task.

Being without field equations the theory does not predict the mean density of matter in space, but it may be possible to obtain that from considerations concerning the formation of condensations.

Finally, an important difference between theories of continual creation (like the steady-state theory; others will be discussed later) and theories of creation in the past only (Newtonian cosmology, general relativity, kinematic relativity) must be pointed out. A theory of continual creation necessarily answers more questions, since no events in the past are required that have no counterpart now. The hypothetical character of continual creation has been pointed out, but why is it more of a hypothesis to say that creation is taking place now than that it took place in the past? On the contrary, the hypothesis of continual creation is more fertile in that it answers more questions and yields more results, and results that are, at least in principle, observable. To push the entire question of creation into the past is to restrict science to a discussion of what happened after creation while forbidding it to examine creation itself. This is a counsel of despair to be taken only if everything else fails.

12.7. A somewhat different approach has been proposed by Hoyle (1948, 1949). He fully accepts the general picture of the universe given by the steady-state theory including necessarily the process of continual creation. He considers it preferable, however, to base the model not on an overall principle like the perfect cosmological principle but on a suitable modification of the field equations of general relativity. He has thus found a set of local physical laws which has the steady-state universe as a consequence. Hoyle's theory is hence closer to the theories that start from locally established physical laws rather than from a cosmological principle. He strongly stresses (1949) that it is the more traditional and well-established procedure of physics to postulate laws that link physical quantities at the same point, and to work out the consequences of these laws for the entire system rather than to postulate general properties of the system as a whole. He also points out that the fact that the cosmological principle does not apply in detail must gravely weaken confidence in its validity.

The theory branches off from orthodox general relativity at the point where that theory generalizes the locally established field equations

$$-\kappa T_{\mu\nu} = R_{\mu\nu} - \tfrac{1}{2}Rg_{\mu\nu}, \tag{12.7}$$

by including the cosmological term $\lambda g_{\mu\nu}$ on the right-hand side (cf. p. 96). It will be remembered that this term was added for purely cosmological reasons which later underwent an alteration, and that the choice of this particular term was dictated by four *desiderata*:

(i) The tensor character of (12.7) was to be preserved, i.e. the equation was still to be valid irrespective of the state of motion of the observer.

(ii) The conservation of energy $(T^{\mu\nu})_{\nu}$ was to apply to the generalized law.

(iii) The new term should be negligibly small in local applications but of great importance on the cosmic scale.

(iv) Simplicity.

The difficulties of relativistic cosmology have been fully discussed in Chapter x. They are connected with the problem of the time-scale and with the hypothetical high densities in the past. Continual creation is a way out of these difficulties, and Hoyle shows that it is possible to make changes in general relativity so as to incorporate this concept. This implies of course that (ii) is to be dropped. But Hoyle points out that there is very little basis for (i) as well. Several years after Einstein had introduced the λ term, Weyl formulated the principle which bears his name, according to which there is a preferred velocity at every point of space-time (cf. p. 100). Not all observers, but only certain fundamental observers, see the universe in the same way. In orthodox relativistic cosmology account is taken of Weyl's principle only when the field equations are solved. But when the purely cosmological term $\lambda g_{\mu\nu}$ (which is only used for these solutions) is introduced, it is selected by prescribing general invariance and hence by explicitly denying Weyl's principle at this stage.

If it is necessary to modify the locally established field equations in order to apply them to the cosmological problem, then that modification should be in accordance with the recognized fundamental ideas of cosmology such as Weyl's postulate. Hoyle's modification of (12.7) is firmly based on this postulate and is as follows. It will be remembered that Weyl stated that the world lines of the nebulae formed a bundle of geodesics diverging from an event O in the (finite or infinitely distant) past. The geodesic joining any event P with O will in general be uniquely defined. Let C be a scalar function of position defined to be proportional to the geodesic distance OP. (Should O turn out to be an infinite distance from all relevant P, C may be defined so that the difference between C at two points P, Q is proportional to the (finite) difference between the geodesic distances of P and Q from O.) In a smooth substratum C is of course the cosmic time, but Hoyle's definition holds also in the presence of condensations.* The first derivative C_μ of C defines a universal field of vectors of constant length pointing, at P, along the geodesic OP produced. The second derivative defines a symmetric tensor field $C_{\mu\nu}$. The modified field equations are

$$-\kappa T_{\mu\nu} = R_{\mu\nu} - \tfrac{1}{2} R g_{\mu\nu} + C_{\mu\nu}. \qquad (12.8)$$

It can be shown that (12.8) possesses the de Sitter metric

$$ds^2 = dt^2 - (dr^2 + r^2 d\theta^2 + r^2 \sin^2\theta \, d\phi^2) \exp(2t/T) \qquad (12.9)$$

as a solution if the universal length of C_μ is $3/T$. Moreover, it can be shown that the solution (12.9) is stable, and in fact that any other solution tends to (12.9) asymptotically.

The density of matter in space is not zero for this metric as in relativistic cosmology but is given by

$$4\pi\gamma\rho T^2 = 1{\cdot}5. \qquad (12.10)$$

* In Hoyle's second paper (1949) the C-field is defined in a slightly different way.

The term $C_{\mu\nu}$ is of the same order of magnitude as the cosmological term in relativistic cosmology, and is hence of no importance when local dense concentrations of matter are considered. Therefore (iii) is satisfied and the simple definition of the C-field satisfies (iv).

Energy (matter) is not conserved since

$$-\kappa(T^{\mu\nu})_{\nu} = (C^{\mu\nu})_{\nu} \neq 0. \qquad (12.11)$$

That this non-conservation is in fact a continual creation of matter is clear, since (for positive T) the de Sitter metric gives an *expanding* universe, but by (12.10) the density remains constant.

The solution (12.9) is, of course, only one of the many possible solutions of (12.8). It represents the smoothed-out substratum. It is evident that its general properties will be exactly like those of the substratum of the steady-state theory. Just as in that theory, so in this one, no difficulties of time-scale arise. The thermodynamic state of the universe is fully explained, the total observable energy and entropy are conserved, and so on. The chief difference is that Hoyle's theory has a perfectly definite set of equations valid both for the substratum and for the actual universe with its non-homogeneities, whereas the theory due to Bondi and Gold does not yield such a set of equations.

However, Hoyle's equations become exceedingly complicated as soon as actual non-homogeneities are introduced, since the entire past history is required in order to construct the field of the C scalar. As Hoyle has pointed out, the stability of the system is so great that for any desired degree of accuracy it is sufficient to go only a finite way into the past since the perturbations of the C-field due to inhomogeneities quickly die out. Similar arguments show that the multiplicity of values for C due to the occurrence of intersecting bundles of geodesics is of little importance. In spite of this interesting and highly significant stability, Hoyle's equation cannot be called useful in practice.

Whether it is preferable to deduce this type of universe from a universal principle (perfect cosmological principle) or from a small modification of locally well-substantiated equations, as in Hoyle's theory, is largely a matter of taste. However, it is of importance to notice that Hoyle's theory gives a perfectly definite value for the mean density of the universe (about 10^{-29} g./cm.^{3}), whereas in the theory of Bondi and Gold no such statement is made.

Hoyle stresses that his theory gives substance to Weyl's principle, whereas its status in relativistic cosmology is rather indefinite. In relativistic cosmology it cannot be a law of nature, since it does not enter the equations, nor can it be an overriding principle, since it does not seem to be true in detail in the actual universe. In Hoyle's theory it enters the equations in a perfectly definite way, laying down, as it were, the rails on which the universe runs.

An interesting problem in this theory is concerned with the formation of condensations. The perfectly definite laws of motion and prescribed mean density imply that this is a purely mathematical problem which should throw light both on the process of nebular condensation and on the applicability of the theory.

CHAPTER XIII

THE THEORIES OF EDDINGTON, DIRAC AND JORDAN

13.1. In the preceding chapters theories have been discussed that start either from the most widely accepted formulation of macrophysics (general relativity) or from the cosmological principle. In neither case is there an explicit denial of the existence of a link between cosmology and microphysics, but the theoretical explanation of the link suggested by the numerical coincidences (Chapter VIII) is left for the future. Doubts may well be entertained whether this is the correct procedure. The numerical coincidences are striking numerical facts directly presenting themselves from observations of physical constants, and there is no obvious reason why they should be considered less primitive than, say, gravitation. The chief aim of the theories, which will be briefly described in this chapter, is to take up the challenge presented by the numerical coincidences and to make their explanation a principal aim of cosmology.

In fact, none of these theories has gained widespread support. Whether this is due merely to the strangeness of their approaches to the problem, or whether this is a consequence of various minor assumptions made for the sake of progress, or whether it is due to direct conflict with observation is an open question, at least in the theories of Eddington and Dirac. Jordan's theory seems to lead to fairly definite contradictions with observation.

13.2. Eddington (1936, 1946) was greatly impressed by the fact that although Dirac had been able to unify special relativity and quantum theory, the union of general relativity with quantum theory seemed to be so very difficult to achieve. He formed the opinion that the difficulty was connected with the deep problems of the particle structure of matter, and that only a searching analysis could show the way. The analysis that he developed to deal with this problem is so very complex and based on such extremely difficult arguments that most scientists found themselves unable to examine the theory in detail. The wealth of the startling numerical results of his theory which offered explanations for most of the pure numbers revealed by observation seemed to act mainly as a deterrent to the study of his work. A deductive theory that aims at showing that an observationally known fact is a necessary consequence of the process of human thought is immediately suspected of fitting the means to the end. The involved nature of Eddington's arguments did little to lessen this suspicion. If a theory of this type makes a correct forecast (as, for example, was the case with general relativity and the bending of light rays by the sun), a general acceptance of the theory is made much more likely. Eddington's work did not make any such forecast, but exclusively explained the known values of the observational numerical constants. A small but fairly well-established discrepancy between the observational and theoretical values of the fine-structure constant lessened faith in the theory still more. The result of this was that only a very few mathematicians made the determined effort required to follow Eddington's technique and arguments, and some of them developed different and less cosmologically-based structures (Bastin and Kilmister, 1952–8). The present author has to confess that he does not yet belong to the small group that has studied Eddington's work in sufficient detail to pronounce on its merits. Accordingly, only the briefest description can be given here, limited as it must be in any case by the complex mathematical apparatus of the theory. For any fuller information, reference must be made to Eddington's two books devoted to the subject.

In attempting to combine general relativity and quantum theory Eddington considered as especially important the introduction of the uncertainty principle into the definition of the reference frames, and the examination of the local observational effects of the universe, particularly those due to the finite number of particles and the finite volume characteristic of the Einstein state. In order to deal with these questions a new mathematical discipline, the wave-tensor calculus, is developed. The careful consideration of the local effects of the structure of the universe is equivalent in many ways to a theory of Mach's principle. The analysis of the phase space constantly throws up the factors 136 and 137 which are closely related to the atomic constant $\hbar c/e^2$.

The general behaviour of the universe is taken to be described by the Eddington-Lemaître model which has been discussed in Chapters IX and X. Although the radius and density of this model vary in time, the data corresponding to its original Einstein state are taken to be characteristic of it. They are closely related to the cosmical constant λ.

Neither the problem of condensation nor the questions concerning the Eddington-Lemaître model mentioned in § 10.6 are discussed.

The chief results of the theory are a number of numerical relations between the constants of nature, relations which, though complicated in derivation, are in general in excellent agreement with observations. Most of these concern only atomic constants, but one relates λ to them. In the Eddington-Lemaître model the constant of the velocity-distance relation varies in time, rising from zero at $t = -\infty$ (Einstein state) to the value $(\frac{1}{3}\lambda)^{\frac{1}{2}}$ at $t = +\infty$ (de Sitter state). Eddington evaluated $(\frac{1}{3}\lambda)^{\frac{1}{2}}$ from atomic constants, arriving at 572·4 km./sec./megaparsec, close to Hubble's value of 540 km./sec./megaparsec. Slater (1947, 1954) found an error in Eddington's calculation and arrived at a corrected figure of 260 km./sec./megaparsec. Though still much higher than recent observational determinations, this discrepancy would merely seem to indicate that the universe is much further from the eventual de Sitter state than Eddington thought when he interpreted his results, and need not be taken as a disproof of his theory.

In this theory space is finite. The number N of elementary particles is quite definite and turns out to be

$$N = \tfrac{3}{2} \times 136 \times 2^{256} \doteqdot 2{\cdot}4 \times 10^{79}.$$

The ratios discussed in Chapter VIII are closely related to the square root of this number.

13.3. It is clear that only a highly complex mathematical theory can be expected to yield numbers of the order of magnitude of N or $N^{\frac{1}{2}}$. The nature of the wave-tensor calculus is sometimes considered to indicate its artificiality. If, therefore, its validity is denied and any similar method ruled out, then one is forced to deny that these enormous numbers are of fundamental significance. Hence their values (though not their coincidences) must be ascribed to an accidental factor entering all the observations. The only such factor that is at all apparent is the timing of the observations, that is, the particular value of the present epoch. The assumption can then be made that all the very large pure numbers of physics do not have any intrinsic significance, but are functions of the epoch having their particular values at the present time only. They are not to be considered as constants, and no special meaning can be attached to their present values. This line of argument is due to Dirac (1937). It clearly differs greatly from all other theories. It wholly denies the validity of Eddington's aim of showing that the known values of the large numbers were the only possible ones. It is contraposed to the basic arguments of the steady-state theory, since it supposes that not only the universe changes but with it the constants of atomic physics. In some ways it may almost be said to strengthen the steady-state arguments by showing how limitless the variations are that may be imagined to arise in a changing universe. Dirac's argument also contrasts sharply with kinematic relativity in that it emphasizes the cosmological significance of atomic dimensional standards, whereas Milne's dimensional hypothesis explicitly denies any such possibility.

In some ways Dirac's argument may be called a counsel of despair. It is due to the failure of other theories to suggest any acceptable argument that might lead to the explanation of the very large numbers. Continued persistent failure to give reasons for the magnitude of these numbers would certainly strengthen support for it.

Dirac's argument, in spite of its great significance, offers no very clear guidance for the development of a cosmology in the way that, say, the perfect cosmological principle does, though like that principle it immediately suggests that current macrophysical theories are inadmissible in cosmological applications.

By using additional assumptions progress may be made, but certainly not all paths have been explored. In fact, only two theories have been developed from the argument. One is a rather brief theory by Dirac himself, while the other one, much more lengthy but almost certainly false, is due to Jordan. It is important to keep in mind that although both theories have found little support this does not necessarily imply that Dirac's argument on its own merits can be rejected.

13.4. Dirac, in his paper, considers the ratio of the reciprocal of Hubble's constant ($1{\cdot}8 \times 10^9$ years) to an atomic unit of time ($e^2/mc^3 = 9{\cdot}5 \times 10^{-24}$ sec.) to be of the order of the age of the universe in natural units. Let us call this ratio $p = 6 \times 10^{38}$. Dirac points out that by choosing any other atomic unit of time one would have arrived at a value of the ratio differing by at most a few powers of ten from this enormous number. He then states his principle in the form: Any two of the very large dimensionless numbers occurring in nature are connected by a simple mathematical relation in which the coefficients are of order of magnitude unity. This is a somewhat stronger formulation than the argument mentioned in § 13.3.

The average density of matter in space expressed in protons per unit atomic volume $(e^2/mc^2)^3$ turns out to be a number of order p^{-1}. This is equivalent to the two statements in Chapter VII that

$$e^2/\gamma m^2 \doteqdot p, \quad \gamma \rho_0 T^2 \doteqdot 1.$$

Using, as in Chapters IX and X, $R(t)$ as the scale factor of the expansion, and assuming mass to be conserved ($\rho R^3 = \text{const.}$), the identification of the density and Hubble's constant ($\dot{R}/R$) leads to the relation

$$\rho_0 \sim R^{-3}(t) \sim p^{-1} \sim R/R. \tag{13.1}$$

It follows that

$$R(t) \sim t^{\frac{1}{3}}, \tag{13.2}$$

and

$$\frac{\dot{R}}{R} = \frac{1}{3t}. \tag{13.3}$$

Hence the age of the universe in these atomic units is only a third of the reciprocal of Hubble's constant, i.e. only 4×10^9 years. In order not to lead to a contradiction with estimates of the ages of the stars (Chapter VI), it may possibly be necessary to assume that nuclear processes occurred more rapidly compared with atomic processes in the past than now. Dirac does not consider this improbable in such a theory, but does not examine the variation of the rate of nuclear processes.

The spaces $t = \text{const.}$ must be flat in this theory. For any curvature of the three-dimensional space would define a radius of curvature whose length would be proportional to $R(t)$. The product of the cube of the radius of curvature and the mean density of matter would define a mass which would be constant by virtue of the conservation of mass. The number of elementary particles making up this mass would be defined purely by astronomical and atomic measurements. This number would be very large. By virtue of Dirac's principle it would be simply related to the other very large numbers like $e^2/\gamma m_e m_p$. But this is impossible since they are changing, while the number mentioned is constant. This ***reductio ad absurdum*** proves that the 3-spaces t = const. are flat.

Up to this stage of development the theory is free of any logical defects, though a number of assumptions have been made. Dirac's attempt to link it with general relativity is, however, very unconvincing. In the atomic units so far employed e and m are constants. But $e^2/\gamma m^2$ is a number of order p and hence proportional to the time t. Therefore γ must be proportional to t^{-1}. Accordingly, Dirac argues that these atomic units cannot be the units employed in general relativity since there γ is taken to be a constant. Hence a different scale of units must be employed in which γ is constant so that the formulae of general relativity can be used. Again, since the conservation of mass is a cardinal principle of general relativity, the unit of mass cannot be changed. It is then easily shown that the new time measure t^* is

$$t^* = t^2/2t_0, \tag{13.4}$$

and the scale has been chosen so that the rates of passage of the two time-scales agree at $t = t_0$. It can be shown that

$$R^*(t^*) \sim tR(t) \sim t^{\frac{4}{3}} \sim t^{*\frac{2}{3}}. \tag{13.5}$$

The same line of argument as is employed to show that 3-space is flat ($k = 0$) may be used to prove that $\lambda = 0$. Then the formulae (10.14) of general relativity lead to zero pressure and to a positive mass-density of plausible magnitude. The model is in fact identical with 2 (ii) in § 9.6. Since the pressure is observed to be much smaller than the density (cf. (7.4)), this is in good agreement with observation. The great weakness of this line of argument is that since $e^2/\gamma m^2$ is taken to be variable, the charge e must vary even if γ and m are constant. But the conservation of charge forms an integral part of Maxwell's equations, which are at least as fundamental to general relativity as the law of conservation of mass. No theory with variable $e^2/\gamma m^2$ can be reconcilable with the classical formulation of general relativity. The choice made here of taking γ and m constant and e variable is quite arbitrary. Even special relativity is entirely based on the laws of propagation of light which, since they follow from Maxwell's equations, are intimately bound up with the law of conservation of charge.

In spite of this logical weakness the theory is not in obvious conflict with observation. It is true that a nuclear time-scale is required that is very long compared with the atomic one, but although this is not very plausible it is by no means impossible. The dynamic (general relativity) time-scale is $\frac{2}{3}$ of $R/\dot{R}$ and is hence about 9×10^9 years. This is rather a brief time-scale, but the determinations of the age of the galaxy are too uncertain to rule out this possibility.

13.5. A second theory of cosmology based on Dirac's argument has been developed by Jordan (1947). His theory is developed in far more detail, but much of it seems to be very arbitrary and its results appear to clash with observation.

Jordan avoids some of the difficulties of Dirac's theory by abandoning the law of conservation of mass. Although there is no *a priori* reason whatever for objecting to this step, it is somewhat disconcerting to find that it is taken in consequence of a misunderstanding of Olbers' paradox. For Jordan believes that it is necessary to introduce a positive curvature of space (spherical space) in order to avoid the great brightness of the background light of the sky deduced by Olbers. This matter was fully discussed in Chapter III, where it was proved that the curvature is completely irrelevant to the issue.

In order to avoid any conflict with the principle of conservation of energy Jordan assumes that the new matter enters the universe in so highly condensed a form that the negative gravitational energy balances the energy of matter. He furthermore argues that these highly condensed new accumulations of matter would develop in such a way as to present first the spectacle of a supernova and then turn into an ordinary star. The mass of these condensations is found by the following curious consideration.

Jordan observes that there is a fairly definite upper limit to stellar masses (about 50 times the solar mass), and that these enormous stars contain about 10^{59} proton masses. He interprets this number as the $\frac{3}{2}$ power of the age of the universe in Dirac's units ($p^{\frac{3}{2}}$). He considers, therefore, that the upper limit of stellar masses is due to the operation of cosmological rather than astrophysical laws, although the latter are commonly believed to supply excellent reasons for the observed mass-distribution.

However, Jordan not only rejects these considerations but also sweeps aside the obvious explanation of the origin of the stars (gravitational condensation). He takes the 'supernova origin' of all (or at any rate most) stars to be the correct explanation. The number of stars per galaxy and the number of galaxies in the universe are both interpreted as being proportional to $p^{\frac{1}{4}}$. The average rate of occurrence of supernovae can be deduced from this and turns out to be a little more than one per galaxy per year. This is in flagrant contradiction with the observed rate (one per galaxy per 200 or 300 years). Jordan's 'explanation' of the contradiction, that the observed rate applies only to 'old galaxies' while the rate is even higher than the average in 'new' galaxies, is valueless; for the brightness of 'new' galaxies so deduced would enable them to be observed throughout the universe at frequent intervals (once a year or so). They would either be exceedingly bright or last a considerable fraction of a year. Hence they could not have escaped observation and thus the theory must be false.

The mathematical formulation of the theory is based on the generalization of relativity known as projective relativity, in which a variable constant of gravitation plays a large part.

The more recent formulation of the theory (P. Jordan, 'Schwerkraft und Weltall', in Series *Die Wissenschaft*, Vol. 107; Braunschweig: Friedr. Vieweg & Sohn, 1955) represents a substantial revision and the criticisms given above do not necessarily apply. However, the author of this book has not made a close study of this version of the theory.

CHAPTER XIV

THE PRESENT POSITION IN COSMOLOGY

14.1. The chief purpose of this chapter is to summarize the changes that have taken place in cosmology since the first edition went to press in 1950, and to indicate briefly directions for observational and theoretical research that seem to be particularly likely to yield important results.

The dominating feature of recent observational work has undoubtedly been the revision of the distance scale, and with it of Hubble's constant, by Baade and Sandage. It is not easy to appreciate now the extent to which for more than fifteen years all work in cosmology was affected and indeed oppressed by the short value of T ($1{\cdot}8 \times 10^9$ years) so confidently claimed to have been established observationally. The time-scale difficulty, as the discrepancy between T and the ages of the Earth and stars was called, exerted a powerful influence on the formulation and development of cosmological theories, and the effects of the removal of this influence have not yet been worked out fully. Hence a discussion here may be appropriate.

In relativistic cosmology the time-scale difficulty has been one of the principal reasons for preferring Lemaître's model to all others. Of course many scientists, notably Lemaître himself, feel that this model has so many other great advantages, such as its finite extent, the definite place given to the formation of galaxies, and the nature of the initial stage, that they do not consider it worthwhile to examine other relativistic models in detail. However, these reasons are not necessarily decisive, and some other models would appear to merit detailed consideration. Some scientists are particularly interested in models of Class V (p. 86), i.e. cases 1 (iii), 2 (iii), 3 (iii*a*), 3 (iv) (pp. 81–5). These are models that start from a point, expand to a finite size and then contract again to a point. The point is characteristic of Newtonian theory as presented in Chapter IX, and of relativistic theory without pressure. If pressure effects are allowed for, by some equation of state, at least in the stages of high density, then the relativistic model does not contract to a point but reaches a minimum radius and then expands again. The same result can be achieved both in Newtonian and in relativistic theory by allowing for local rotations and for anisotropy (Zelmanov, 1955; Raychaudhuri, 1956; Heckmann and Schücking, 1958). Such a model, then, oscillates for all time between a maximum and a minimum size, having alternate expanding and contracting phases. The present would clearly be identified with an instant during an expanding phase.

Such a model has the advantage of being unending in time, like the steady-state model, but without the need to invoke continual creation. The system ages during each expanding phase and rejuvenates during each contracting phase. Just how this rejuvenation process would work depends on the thermodynamics of the contracting phase which has not yet been worked out in sufficient detail. Both Olbers' paradox and the Wheeler–Feynman absorber theory of radiation require reconsideration during contraction when conditions would be almost unimaginably different from the present.

Another model that deserves attention now is that of Einstein and de Sitter in which $k=0$ and $\lambda=0$ so that $R(t)\sim t^{\frac{2}{3}}$ (p. 82). In addition to its outstanding simplicity, this model has the remarkable property (unique amongst relativistic models) that $\gamma\rho R^2/\dot{R}^2$ is constant. It may well be argued that as important a simple pure number as $\gamma\rho R^2/\dot{R}^2$ should be constant during the evolution of the universe so as to provide in some sense a constant background to the application of the theory.

Since the death of Milne very little work has been done on kinematic relativity and so it is not clear how the change in Hubble's constant affects this theory.

Although the steady-state theory did not suffer directly from the time-scale difficulty it would have been scarcely possible within the theory to reconcile the old time-scale with modern ideas of galactic evolution, the average age of a galaxy in the theory being $\frac{1}{3}T$. With $T=1{\cdot}8\times10^9$ years this would give an average age of a galaxy of 6×10^8 years, which is far lower than is inferred by other lines of thought. With the modern value of T, an average age of 4×10^9 years is obtained, which is very plausible.

14.2. The chief dichotomy in theoretical cosmology appears to lie between the steady-state and evolutionary theories, and so an observational decision on this point would seem to be particularly desirable. Fortunately there are numerous ways in which relevant observations appear to be possible with existing observational techniques, and some others in which a development of astrophysical theories might lead to decisive results. It will be realized that in opposing one very inflexible theory, the steady-state theory, to *all* the evolutionary models, an observational result can far more easily disprove the steady-state theory than *all* the other theories. In fact, since the steady-state forecasts a perfectly definite answer to each of the observations enumerated, while the others forecast a whole range of answers, one would expect almost every observation to disagree with the steady-state theory at the limit of observation, where errors are always large, even if the theory were in fact correct. Since observational errors are frequently under-estimated, as experience shows, one would expect, if the theory were correct, to meet claims that the steady-state theory had been disproved just at the very limits of the observation each time a new type of observation has been made, and for the claim to be withdrawn after a more critical analysis of the errors involved. In fact this has happened more than once (pp. 43–4) in the last ten years.

The most hopeful lines of attack on the cosmological problem appear to be the following:

(i) Radio-astronomical number counts: Further work in this promising field may be decisive, particularly if the distance of a few more radio sources can be established and so the cosmological relevance of the data confirmed.

(ii) Optical number counts: In view of the difficulty of measuring luminosities of diffuse objects and of the uncertain effect of the K-term such measurements could be convincing.

(iii) Measurements of angular diameters of very distant objects: According to the steady-state theory these decrease monotonically with distance to a finite limit; according to almost all other theories (and neglecting evolutionary changes) they diminish with distance to a minimum and then increase again. Only radio observations are likely to be able to cover sufficient distances to show this effect.

(iv) Variation of physical characteristics of galaxies with distance: This seems to be a most important line of approach. If the universe is evolving, then distant galaxies should be seen to be at a lower age than near galaxies, while in a steady-state there should be no such variation on an average. There are numerous possibilities of investigating this effect. While the presence of any such variation would tell decisively against the steady-state theory, it is difficult to say out to what distance the absence of such an effect would have to be observed in order to disprove evolution of the universe.

Possibilities of particular interest are as follows:

(*a*) Variation of colour with distance (Stebbins–Whitford effect): As was mentioned earlier, it was thought in 1948 that such a variation had been discovered in the range out to $z = 0{\cdot}03$, but this was disproved later by Whitford and Code. It would obviously be of the greatest interest to extend these measurements to larger distances, though a continued negative result might not be of significance unless $z = 0{\cdot}30$ or so were reached.

(*b*) Variation of shape and of frequency of type with distance: It might not be too difficult to make such measurements, but the interpretation would depend on an accurate assessment of selection effects, which might well be impossible.

(*c*) Variation of degree of clustering with distance: The same comments apply.

(*d*) Variation in the ratio of young stars to old stars: This is a very valuable approach, as is shown by the work of Morgan (1958) on the existence of young-star-rich and young-star-poor galaxies. Again, greater distances than so far reached are required if a positive effect is to be excluded.

(v) Exact red shift-apparent luminosity measurements: Determination of the higher terms of the expansion would be of great interest. However the uncertainties of operating at extreme range, and the selection effects likely to be encountered, do not make this approach as hopeful as might appear.

(vi) Determination of random velocities of galaxies: Since in an expanding universe random velocities between gravitationally independent objects such as clusters decrease in the course of time, in an evolutionary universe these random velocities at the time of condensation of the galaxies were presumably considerably higher than now. However, condensation cannot take place with high peculiar velocities, and so Lemaître's model requires present random velocities of clusters to be no more than about 100 km./sec. Determination of these by optical means is almost impossible, but 21 cm. radio observations might give velocities in the inter-galactic medium sufficiently well to enable this quantity to be determined.

(vii) Theory of galactic evolution: The steady-state theory requires the present population to represent all ages of galaxies in a definite frequency distribution. Whether in fact the galaxies near us (say within 50 million light years) can be said to have different ages depends on the relation between observable characteristics (type, etc.) and age, which theory should be able to elucidate before long. Again in this connection the measurements of Morgan may prove to be of vital importance.

(viii) Theory of the condensation of the nebulae: Much further work needs to be done before the pioneer theory of Sciama (1953), or an alternative one, is sufficiently well established to show whether the conditions for galactic condensation would be realized in an evolutionary or in a stationary universe. The work of Ebert (1955), Bonnor (1956, 1957, 1958) and McCrea (1958) on the effect of pressure on gravitational condensation has brought the solution of this problem nearer.

(ix) Measurements of intergalactic density: This has a direct bearing on (viii), and is immediately related to the conditions that must exist in intergalactic space according to the steady-state theory. Galactic haloes, the observations of bridges between galaxies by Zwicky (1953), studies of the physical state of the intergalactic medium by Field (1958), and suitable 21 cm. line measurements all contribute to our knowledge of the intergalactic medium.

This list is by no means exhaustive and no doubt new evidence will be forthcoming during the next decade.

14.3. If the observations eventually decide in favour of the steady-state theory, then the next task will be for theoreticians to build a firm connection between the ideas of continual creation and those of the rest of physics. Until this is done, the relation between cosmology and, in particular, the theory of elementary particles will be tenuous in the extreme.

If the observational tests decide in favour of evolution, then the next step will have to be the selection of a particular evolutionary model. Several of the tests listed, notably (i), (ii), (v) and (viii), are suitable for this purpose whereas others are not. Reliable determination of the mean density of the universe will also be of assistance in the solution of this question. It will be particularly valuable if the physical content could be worked out for relativistic models other than Lemaître's.

14.4. It must again be emphasized that the tentative suggestions made here can, by their very nature, only concern quantitative improvements of existing measurements. A great qualitative step forward in any branch of physics may have profound effects on cosmology that cannot possibly be forecast now. There are probably few features of theoretical cosmology that could not be completely upset and rendered useless by new observational discoveries. But it is not likely to be too hopeful a view to say that there is a fairly large body of observational cosmology that is more or less secure, and that theoretical cosmology is on the way to becoming an interesting branch of physics, capable of exploration by means of the commonly accepted principles of scientific inference. It would be particularly interesting if cosmology, as a large-scale testing ground of physics, were to acquire an influence on the development of local physical theories.

BIBLIOGRAPHY

The list of books and papers given here is intended to serve as a guide to further study. Accordingly the list is not confined to references given in the text. The works are arranged as far as possible by the chapters of the book to which they refer particularly.

PART 1

CHAPTER I

(1) OLBERS (1826). *Bode's Jahrbuch*, 110.
MILNE, E. A. (1935). See reference (82).

CHAPTER II

(2) BONDI, H. and GOLD, T. (1948). *Mon. Not. R. Astr. Soc.* **108**, 252.
(3) WALKER, A. G. (1944). *J. Lond. Math. Soc.* **19**, 219, 227; *Observatory*, **65**, 242.
(4) WHITROW, G. J. (1936). *Z. Astrophys.* **12**, 47.
WHITROW, G. J. (1937). *Z. Astrophys.* **13**, 113.

For a general discussion of the subject of Chapters I and II, see references (46), (49), (50), (55), (58), (82), (83), (84), (89), (90).

PART 2

CHAPTER III

OLBERS. See reference (1).
GOLD, T. *Proc. Int. Solvay Conference*, 1958. Stoops: Brussels, 1959.

CHAPTER IV

(5) MACH, E. (1893). *The Science of Mechanics*, pp. 229–38. London.
(6) EINSTEIN, A. (1917). *S.B. preuss. Akad. Wiss.* 142.
HOYLE, F. (1948). See reference (89).

CHAPTER V

The observations are described in:

(7) HUBBLE, E. (1936). *The Realm of the Nebulae*. Oxford.
(8) HUBBLE, E. (1937). *The Observational Approach to Cosmology*. Oxford.

The parts of these two otherwise admirable books dealing with the comparisons between theory and observation are open to objections (see p. 115 of this book). Both books contain full references. A few of the principal papers are:

(9) SLIPHER, V. M. (1915). *Pop. Astr.* **23**, 21.
(10) SHAPLEY, H. and AMES, A. (1932). *Ann. Harv. Coll. Obs.* **88**, no. 2.

(11) SHAPLEY, H. (1933). *Proc. Nat. Acad. Sci., Wash.*, **19**, 591.
(12) HUMASON, M. (1931). *Astrophys. J.* **74**, 35.
HUMASON, M. (1936). *Astrophys. J.* **83**, 10.
(13) HUBBLE, E. (1936). *Astrophys. J.* **84**, 158, 270, 517.

References (10) and (11) give an estimate of the mean density of matter. Reference (13) contains a discussion of the K-term.

(14) GREENSTEIN, G. L. (1938). *Astrophys. J.* **88**, 605. Considers the effective temperature of nebulae.
(15) SMITH, SINCLAIR (1936). *Astrophys. J.* **83**, 23.

The estimates of the mass of non-luminous matter are based on reference (15) and on the series of papers by F. Hoyle and R. A. Lyttleton.

(16) HOYLE, F. and LYTTLETON, R. A. (1939). *Proc. Camb. Phil. Soc.* **35**, 405, 592.
(17) HOYLE, F. and LYTTLETON, R. A. (1940). *Proc. Camb. Phil. Soc.* **36**, 424.
(18) HOYLE, F. and LYTTLETON, R. A. (1941). *Mon. Not. R. Astr. Soc.* **101**, 227.
(19) HOYLE, F. (1945). *Mon. Not. R. Astr. Soc.* **105**, 287.

The reddening of the continuous spectra of nebulae is described in:

(20) STEBBINS, J. and WHITFORD, A. E. (1948). *Astrophys. J.* **108**, 413.

Recent papers on the red shift include the contributions by W. Baade and A. R. Sandage to the *Int. Solvay Conference*, 1958 and

HUMASON, M., MAYALL, N. U. and SANDAGE, A. R. (1956). *A.J.* **61**, 97.

The following papers criticize and the last two correct (20):

DE VAUCOULEURS, A. (1953). *Mon. Not. R. Astr. Soc.* **113**, 134.
BONDI, H., GOLD, T. and SCIAMA, D. W. (1954). *Astrophys. J.* **120**, 597.
WHITFORD, A. E. (1954). *Astrophys. J.* **120**, 599.
WHITFORD, A. E. (1956). *A.J.* **61**, 353; see also
CODE, A. D. (1959). *Publ. Astr. Soc. Pacif.* **71**, 118.

Statistical problems are discussed in many papers by Neyman, Scott and Shane in particular:

NEYMAN, J., SCOTT, E. L. and SHANE, C. D. (1953). *Astrophys. J.* **117**, 92; **120**, 606.
NEYMAN, J. and SCOTT, E. L. (1955). *Astrophys. J.* **60**, 33.
SCOTT, E. L. (1957). *Astrophys. J.* **62**, 248.

The cosmological aspects of radio astronomy are discussed in:

GOLD, T. *Proc. Conf. on Dynamics of Ionized Media.* Dept. of Physics, University College: London, 1951.
RYLE, M. *Proc. Conf. on Dynamics of Ionized Media.* Dept. of Physics, University College: London, 1951.

RYLE, M. (1958). *Proc. Roy. Soc.* A, **248**, 289.

RYLE, M., SHAKESHAFT, J. R., BALDWIN, J. E., ELSMORE, B., THOMSON, J. H. (1955). *Mem. R. Astr. Soc.* **67**, 106.

MILLS, B. Y. and SLEE, O. B. (1957). *Aust. J. Phys.* **10**, 162.

(Many of the great contributions of the Sydney group will be found in this journal.)

EDGE, D. O., SCHEUER, P. A. G. and SHAKESHAFT, J. R. (1958). *Mon. Not. R. Astr. Soc.* **118**, 183.

LILLEY, A. E. and MCCLAIN, E. F. (1956). *Astrophys. J.* **123**, 172.

This refers to Bok's earlier work which yielded the current time-scale of the galaxy and upset the conclusions of:

CHAPTER VI

The age of the Earth is discussed in:

(21) JEFFREYS, H. (1929). *The Earth*, p. 64. Cambridge.

(22) JEFFREYS, H. (1948). *Nature, Lond.*, **162**, 822.

(23) JEFFREYS, H. (1949). *Nature, Lond.*, **164**, 1046.

(24) HOLMES, A. (1949). *Nature, Lond.*, **163**, 453.

(25) BULLARD, E. C. and STANLEY, J. P. (1949). *Veröff. finn. geod. Inst.* no. 36, p. 33.

The ages of meteorites are examined in:

(26) PANETH, F. A. (1939). *Occ. Notes R. Astr. Soc.* no. 5, p. 57.

(27) ARROL, W. J., JACOBI, R. B. and PANETH, F. A. (1942). *Nature, Lond.*, **149**, 235.

(28) BAUER, C. A. (1947). *Phys. Rev.* **72**, 354.

The age of the stars is discussed in:

(29) HOYLE, F. (1947). *Mon. Not. R. Astr. Soc.* **107**, 334.

Dynamical arguments occupy the main part of:

(30) BOK, B. J. (1946). *Mon. Not. R. Astr. Soc.* **106**, 61.

For earlier estimates see:

(31) JEANS, J. H. (1928). *Astronomy and Cosmogony*. Cambridge.

The composition of the stars is discussed in:

(32) RUSSELL, H. N. (1933). *Astrophys. J.* **78**, 239.

(33) STRÖMGREN, B. (1940). *Festschrift für E. Strömgren*, p. 218. Copenhagen.

(34) HOYLE, F. (1947). *Mon. Not. R. Astr. Soc.* **106**, 255.

Interstellar material is discussed in references (16)–(20) and in:

(35) DUNHAM, T. Jr. (1939). *Proc. Amer. Phil. Soc.* **81**, 277.

(36) MCKELLAR, A. (1940). *Publ. Astr. Soc. Pacif.* **52**, 187.

(37) BONDI, H., HOYLE, F. and LYTTLETON, R. A. (1947). *Mon. Not. R. Astr. Soc.* **107**, 184.

The origin of the elements is discussed in:

(38) HOYLE, F. (1946). *Mon. Not. R. Astr. Soc.* **106**, 343.
(39) HOYLE, F. (1947). *Proc. Phys. Soc. Lond.* **59**, 972.
(40) ALPHER, R. A. and HERMAN, R. C. (1950). *Rev. Mod. Phys.* **22**, 153.

The last-named paper gives full references to earlier work.

The definitive theory of nucleogenesis is given in:

BURBIDGE, E. M., BURBIDGE, G. R., FOWLER, W. A. and HOYLE, F. (1957). *Rev. Mod. Phys.* **29**, 547.

CHAPTER VII

The numerical coincidences are described in references (93) and (95). Cosmological theories of the origin of cosmic rays are given in references (57) and (82).

PART 3

CHAPTER VIII

(41) ROBERTSON, H. P. (1935). *Astrophys. J.* **82**, 284.
ROBERTSON, H. P. (1936). *Astrophys. J.* **83**, 187, 257.
(42) WALKER, A. G. (1936). *Proc. Lond. Math. Soc.* (2), **42**, 90.

Almost all the books referred to in this part contain some discussion of the subject-matter of Chapter VIII. References (62), (65) and (66) are mentioned in this chapter.

CHAPTER IX

(43) MCCREA, W. H. and MILNE, E. A. (1934). *Quart. J. Math.* (Oxford Ser.), **5**, 73.
(44) MILNE, E. A. (1934). *Quart. J. Math.* (Oxford Ser.), **5**, 64.
(45) HECKMANN, O. (1940). *Nachr. Ges. Wiss. Göttingen*, N.F., **3**, 169.
(46) HECKMANN, O. (1942). *Theorien der Kosmologie*. Springer: Berlin.

The last named is an excellent but almost unobtainable book. It gives a full discussion of the subject-matter of Chapters IX, X and XI, and very full references for the period 1933–40.

(47) NEUMANN, C. (1896). *Über das newtonische Prinzip der Fernwirkung*. Leipzig.
(48) SEELIGER, H. (1895). *Astron. Nachr.* **137**, 129.
SEELIGER, H. (1896). *Münch. Ber. Math. Phys. Kl.* 1896, p. 373.
LAYZER, D. (1954). *Astrophys. J.* **59**, 268.
MCCREA, W. H. (1954). *Astrophys. J.* **60**, 271.

Chapter X

A review of relativistic cosmology with full references is:

(49) Robertson, H. P. (1933). *Rev. Mod. Phys.* **5**, 62.

A brief review of the subject and of kinematic relativity is:

(50) Bondi, H. (1948). *Mon. Not. R. Astr. Soc.* **108**, 104.

The most recent measurements of the Einstein shift are described in:

(51) Adam, M. G. (1948). *Mon. Not. R. Astr. Soc.* **108**, 446.

The following are books on the subject:

(52) Eddington, A. S. (1930). *The Mathematical Theory of Relativity.* Cambridge.
(53) Eddington, A. S. (1933). *The Expanding Universe.* Cambridge.
(54) Tolman, R. C. (1934). *Relativity, Thermodynamics and Cosmology.* Oxford.
(55) Whitrow, G. J. (1950). *The Structure of the Universe.* Hutchinson.
(56) McVittie, G. C. (1937). *Cosmological Theory.* Methuen.
(57) Lemaître, G. (1948). *L'atome primitive.* Genève.
(58) Einstein, A. (1936). *The Theory of Relativity.* Methuen.

References (52) and (58) deal principally with the theory of relativity itself; references (53) and (55) are popular books; a large part of reference (55) and the last part of reference (56) are devoted to kinematic relativity (Chapter xi); references (52), (54) and (56) contain introductions to the tensor calculus. Reference (57) summarizes Lemaître's great contribution. A few of the principal papers on the subject are reference (6), and:

(59) de Sitter, W. (1917). *Mon. Not. R. Astr. Soc.* **78**, 3.
(60) Friedmann, A. (1922). *Z. Phys.* **10**, 377.
(61) Friedmann, A. (1924). *Z. Phys.* **21**, 326.
(62) Weyl, H. (1923). *Phys. Z.* **24**, 230.

The Robertson line element was derived by Robertson (1935), reference (41) and Walker (1936), reference (42).

Papers dealing with the comparison between theory and observation are:

(63) McCrea, W. H. (1935). *Z. Astrophys.* **9**, 290.
(64) McCrea, W. H. (1939). *Z. Astrophys.* **18**, 98.
(65) Hubble, E. and Tolman, R. C. (1935). *Astrophys. J.* **82**, 302.
(66) Hubble, E. (1936). *Astrophys. J.* **84**, 517.
(67) McVittie, G. C. (1939). *Proc. Phys. Soc. Lond.* **51**, 529.
(68) ten Bruggencate, P. (1937). *Naturwissenschaften*, **35**, 561.

(69) Temple, G. (1938). *Proc. Roy. Soc.* A, **168**, 122.
(70) Robertson, H. P. (1937). *Z. Astrophys.* **15**, 69.
(71) Bondi, H. and McVittie, G. C. (1948). *Observatory*, **68**, 111; see also reference (46).
Robertson, H. P. (1955). *Pub. A.S.P.* **67**, 82.

The problem of condensation is discussed in:

(72) Eddington, A. S. (1930). *Mon. Not. R. Astr. Soc.* **90**, 668.
(73) McCrea, W. H. and McVittie, G. C. (1931). *Mon. Not. R. Astr. Soc.* **92**, 7.
(74) Lemaître, G. (1931). *Mon. Not. R. Astr. Soc.* **91**, 490.
(75) McVittie, G. C. (1932). *Mon. Not. R. Astr. Soc.* **92**, 500.
(76) Gamow, G. and Teller, E. (1939). *Phys. Rev.* **55**, 654.

The origin of the elements is linked with the early stages of the universe in many of the papers listed in reference (40) and in:

(77) Gamow, G. (1948). *Phys. Rev.* **74**, 505.
(78) Gamow, G. (1949). *Rev. Mod. Phys.* **21**, 367.

Abandoning the cosmological principle is considered in reference (64) and in:

(79) Eddington, A. S. (1939). *Sci. Progr.* **34**, 225.
(80) Tolman, R. C. (1949). *Rev. Mod. Phys.* **21**, 374.
(81) Omer, G. C. Jr. (1949). *Astrophys. J.* **109**, 164.

Recent work is given in the contributions of G. Lemaître, O. Klein, O. Heckmann, W. W. Morgan and others to the *Proc. Int. Solvay Conference*, 1958. Stoops: Brussels, 1959.

The problem of condensation is discussed by:

Bonnor, W. B. (1956–58). *Mon. Not. R. Astr. Soc.* **116**, 351; **117**, 104; **118**, 523.
McCrea, W. H. (1957). *Mon. Not. R. Astr. Soc.* **117**, 562.
Ebert, R. (1955). *Z. Astrophys.* **37**, 217.
Field G. B. (1958). *Proc. Inst. Radio Engineers* **46**, 240.

Chapter XI

The work on this subject is presented in:

(82) Milne, E. A. (1935). *Relativity, Gravitation and World Structure.* Oxford.
(83) Milne, E. A. (1948). *Kinematic Relativity.* Oxford.
(84) Johnson, Martin (1948). *Time, Knowledge and the Nebulae.* Oxford.

See also references (50) and (55).

The theory is criticized in:

(85) Born, Max (1943). *Experiment and Theory in Physics.* Cambridge.
(86) McVittie, G. C. (1940). *Observatory*, **63**, 273.

The last-named article is replied to in:

(87) MILNE, E. A. and WALKER, A. G. (1940). *Observatory*, **64**, 11, 17.

CHAPTER XII

BONDI, H. and GOLD, T. (1948). See reference (2).
(88) McCREA, W. H. (1950). *Endeavour*, **9**, 3.
(89) HOYLE, F. (1948). *Mon. Not. R. Astr. Soc.* **108**, 372.
(90) HOYLE, F. (1949). *Mon. Not. R. Astr. Soc.* **109**, 365.
McCREA, W. H. (1951). *Proc. Roy. Soc.* A, **206**, 562.
SCIAMA, D. W. (1955). *Mon. Not. R. Astr. Soc.* **115**, 3.

CHAPTER XIII

(91) EDDINGTON, A. S. (1936). *Relativity Theory of Protons and Electrons*. Cambridge.
(92) EDDINGTON, A. S. (1946). *Fundamental Theory*. Cambridge.
BASTIN, E. W. and KILMISTER, C. W. *Proc. Camb. Phil. Soc.* **50**, 278; **51**, 279, and later papers in the same journal.
SLATER, N. B. (1947). *Phil. Mag.* **38**, 299.
SLATER, N. B. (1954). *Nature, Lond.*, **174**, 321.

For a criticism see reference (85).

(93) DIRAC, P. A. M. (1938). *Proc. Roy. Soc.* A, **165**, 199.

An earlier formulation is given in·

(94) DIRAC, P. A. M. (1937). *Nature, Lond.*, **139**, 323.

Jordan's theory is developed in:

(95) JORDAN, P. (1947). *Die Herkunft der Sterne*. Stuttgart.
(96) JORDAN, P. (1949). *Nature, Lond.*, **164**, 637.

INDEX

Bold type indicates the principal references: capitals denote authors' names